THE WORLD COMMUNICATES

The World Communicates

Maurice Rickards

The picture shows a bank of 'monitor' teleprinters at an automatic telex exchange. These are used for checking operational accuracy. International telex connects teleprinters as quickly and simply as the conventional exchange connects the ordinary telephone. Direct dialling gives fully automatic operation. Receiving machines need not necessarily be manned; messages may thus be transmitted forward from the world's 'daytime' areas into the unattended offices of countries where it is already night. The typed message, accurately recorded, is ready for the morning.

A LONGMAN
WORLDBOOK

LONGMAN GROUP LIMITED LONDON

Associated companies, branches and representatives throughout the world

© Maurice Rickards 1972

All rights reserved. No part of this publication may be reproduced, stored in a retrieval system, or transmitted in any form or by any means, electronic, mechanical, photocopying, recording, or otherwise, without the prior permission of the Copyright owner.

Designed by Maurice Rickards

First published 1972

ISBN 0 582 15051 5

Printed in Spain
by Editorial fher s.a. - Bilbao

CONTENTS

RUNNERS, RIDERS AND PIGEONS   7
        The Beginnings of Keeping in Touch

A WORLD OF INSTANT CONTACT   16
        Electricity Leaps the Frontiers

POST-BOX OR HAND-SET?   26
        The Electronic Way from A to B

NEWS CIRCLES THE GLOBE   44
        Speeding the Flow of Words and Pictures

SEEING AT A DISTANCE   52
        New Eyes for Science and Industry

NOW, FACSIMILE AND VIDEO   61
        Vision by Wire, Tape and Disc

MESSAGES THROUGH SPACE   70
        The Role of the Communications Satellite

INTO THE EIGHTIES   79
        New Problems—New Solutions

RUNNERS, RIDERS AND PIGEONS

The Beginnings of Keeping in Touch

THE URGE to communicate is among man's most powerful drives. The fear of fire and pestilence and famine is deep-rooted, but running these a close second is the fear of isolation. By gesture, grunt, discourse, essay or symbol—even by mere proximity—man seeks the comfort of keeping in touch.

This simple proposition has a lot to answer for. It is the mainspring of a hundred thousand human activities, some of them good, some of them bad. Cat-calls, songs, newspapers, tribal warcries, television programmes—all are facets of the same need. The power of speech, itself most distinctive of all outward human characteristics, is the basic symptom; the rest follow naturally.

Speech serves as a pathway between individuals. The emergence of this code of sound, this intricate system of interconnection, was a big step. It marked a major stage in human development. The ability to communicate thoughts transformed Man from a single unit—lively, fearful and aggressive, to a social being—collective, confident and talkative.

As well as being talkative he is inclined to be lazy; one of his earliest concerns was the avoidance of unnecessary effort. The transmission of a message from point A to point B could involve a quite exceptional amount of effort. It meant, unless he could think of some other way, actually taking the message himself—moving from A to B with the message in his head and speaking it aloud on arrival—a tiring and demanding process. There were a number of early solutions to the problem, none of them ideal, all of them calling for a high degree of planning and cooperation.

You could send a message through the air without moving from the spot; you could, for example, *shout* it. Xerxes, King of the Persians, devised a complete communications network on this basis. He set up a human loud-hailer chain—a corps of shouters. Military messages were transmitted by a series of soldiers, each specially trained for maximum volume, each passing on the message to the next man as soon as he heard it. The men were posted on hilltops and used megaphones made from animal hides; often there were hundreds, sometimes thousands of men in the chain. By using this call-post system, the length and breadth of the country could be covered in twenty-four hours.

Beacon fires along the Hudson signalled peace between the Americans and the British in 1783. The hill-top flare, a signal since earliest times, was narrowly limited in message-content and required leisured pre-arrangement.

Ungrammatical and unreliable, news transmission in the early 1800s was also slow. The breathless headlines of this 1813 news sheet suggest messages hot from the scene of battle. In fact the events referred to had occurred over a period of some months, none of them more recently than a week or so before publication.

Europe delivered
The TYRANT'S ARMY Annihilated.

GAZETTE EXTRAORDINARY.

90,000 French Killed, Wounded, and Prisoners in several successive Battles, in which the Talents and Bravery of the Allied Armies were eminently conspicuous.

180 *Pieces of Canon taken,*
Dresden surrendered,
Leipsic taken.

The King of SAXONY and his Family made Prisoners; the Saxon, Baverian, and Wertenburgh Armies all taken.

20 French Generals Killed.
BUONAPARTE RUN AWAY.

Pigeons have served as messengers into recent times—including World Wars I and II. The picture shows a London bus in use as a pigeon loft on the Western Front in World War I. When tanks were introduced in 1917, provision was made for 'pigeon-release ports' for emergency messages.

There were obvious snags. For one thing, calling distances were limited to the throw of the human voice: the system used up a lot of men, most of whom, at any given time, were doing nothing. For another thing, the message at the end of the line was not always the one that started out. At the best of times mis-hearings along the line built up from one post to the next; in windy weather the system went to pieces altogether. But within its limitations, crude as it was, the system worked.

An alternative method—also airborne—was visual. Here the range was much extended; where a voice could be heard only over a few hundred yards, a puff of smoke or a spurt of flame could be seen for many miles. Again, with the post-to-post principle applied, great distances could be covered quickly and relatively cheaply. From the days of the early scriptures, long before the birth of Christ, till as recently as the sixteenth century, when news of the approach of the Armada was flashed from beacon to beacon across Britain, the hilltop flame has served as messenger.

Yet another method—again an alternative to going from A to B yourself—was to send somebody else. This idea, though slower than sound or light, had a number of advantages: the message could be longer and more complex; if rendered as marks on a

suitable material (wax, parchment or paper) it would not deteriorate in transit. In this form it was also private, as it could be sealed or folded so that the messenger himself did not know what he was carrying. One of its most attractive features was that it offered a saving in effort on the part of the sender. After speech, the second most significant breakthrough was the written word.

Sending someone else became standard practice. At first it was just a man on foot; one such bearer of a message, having run twenty-two miles with news of his country's victory over the Persians at Marathon (490 BC) gasped out his message on the steps of the Acropolis in Athens and fell dead. But for the most part journeys were divided into organized sections, with runners taking over from each other at given intervals.

Later came riders—horsemen, charioteers and coachmen—who travelled faster and farther, and who carried the post-to-post system into the days of the pony express, the post horse and the stagecoach. By the middle of the eighteenth century the whole of Europe was criss-crossed with a network of staging posts, each a day's ride from the other, each supplying refreshment and fresh horses for the onward journey.

The Eastern Pony Express, inaugurated in 1836, linked America's

A hundred years ahead of his time, Georges Louis le Sage was an early experimenter in the field of electrical communications. He set up a device consisting of a twenty-four strand 'cable-link', each strand of which ended in a hanging pith ball. With each strand corresponding to a letter of the alphabet, individual wires were touched in sequence with wax rods electrified by friction. Each contact produced a divergence in the pith ball at the other end of the line. This contemporary illustration is intended to show the device in use between two houses. The wall is removed in the drawing to show the principle.

For centuries communications relied on the pack horse, the wagon team and the passenger coach. Roads, some scarcely more than trackways between one point and another, were often impassable except in summer. In his 'Tour through Great Britain' Defoe wrote '... sometimes a whole summer is not dry enough to make the roads passable'. It was still true a hundred years later, when the pack-horse trains and wagoners had given way to the coach. This picture, dated 1842 and showing the road through the pass at Llanberis, conveys something of the sense of loneliness and inadequacy that beset the traveller by coach. The pace was slow; London to York took four days; Paris to Marseilles took upwards of a week and a half. The construction of roads—not to mention the problem of their maintenance—was everywhere a matter of neglect. Not since the time of the Romans had there been a unified and effective system of roadways. Road technology was bad; administration was patchy and corrupt. Many of the 'turnpike' tolls, levied to raise money for upkeep, were diverted to private gain. The roads revolution started with unified control and the development of roadmaking techniques initiated by Tresuguet (France 1764) and Macadam (Britain 1798).

Published during the severe frost of 1854, this three-part picture strip sums up the progress of Britain's postal service. The snowbound postboy and mail coach epitomize the uncertainties of days then only recently gone; the third picture, though the train, too, is snowbound, is the expression of modern times; 'our illustration,' says the caption, 'shows the mail train at a stand-still, blowing off its steam in the face of a heavy drift; the passengers, no doubt, impatient at the delay; yet how comfortable their condition in contrast with that of passengers by the old mail coaches!' As the railways opened up communications they aroused both admiration and hostility. Some observers saw in them liberation —others, desecration.

big eastern cities with St Louis, New Orleans and Charleston, and provided a regular six-and-a-half-day service between New York and New Orleans, a distance of twelve hundred miles. Later, in the early 1860s, the Western Pony Express offered a regular service between New York and San Francisco—two thousand five hundred miles in just over ten days. But by this time, with the era of the railways and telegraphs opening up, the horse had begun to vanish from the picture.

A much earlier precursor of the horse—and one that required no one to go with it—was the pigeon. Known since earliest times for its ability to get home on its own, even from very great distances, the pigeon has never been completely out of the picture. First records of the pigeon as messenger date from the fifth Egyptian dynasty (about 3000 BC). Hirtius, Roman Consul, used them to get messages to Decius Brutus, under siege in Modena. So did the Sultan of Baghdad, to get news from Syria and Egypt. So did the defenders of Paris, to keep in touch with the outside world as the Germans besieged the city in 1870.

The Siege of Paris is a landmark in the history of communications. In the 1860s and early 1870s the railway and the telegraph systems were just getting into their stride, but the stagecoach and the post-chaise were by no means dead. There had even been rumours of aircraft—powered flight by 'heavier-than-air machines', but this was wild talk; Paris still boasted half-a-dozen hot-air balloons from the era of Montgolfier. At this mid-point in history the city was to serve as a focus of the last big balloon operation, as a base for the first big pigeon-post set-piece, and curtain-raiser for the use of photography in document transmission.

There was one further feature that distinguished the siege of Paris: in a world of rapidly developing communications a major city of modern times found itself completely cut off from the outside world. Suddenly, for every inhabitant, there was the unaccustomed fear of isolation.

In his account of the siege, John Fisher says that the hardest part was 'not the famine and the shelling but the isolation, the absence of news, the complete ignorance of what was happening only a few miles away'. He quotes Théophile Gautier, poet and dramatist, one of the two million trapped in the city: 'We were absolutely cut off from the world, not knowing what they did in the provinces, whether or not they were coming to our aid, abandoned by the universe . . .'

On 2nd September 1870 the army of Napoleon III surrendered to the German forces at Sedan. Paris, left to her own resources, stunned by a defeat that was unthinkable, had gathered her strength for the siege that was inevitable. As news came that the last remaining telegraph cable had been cut it was clear that other means must be found to keep contact with the outside world. Not only from the point of view of public morale but in a strictly military sense, it was vital to the defenders of the city to know what

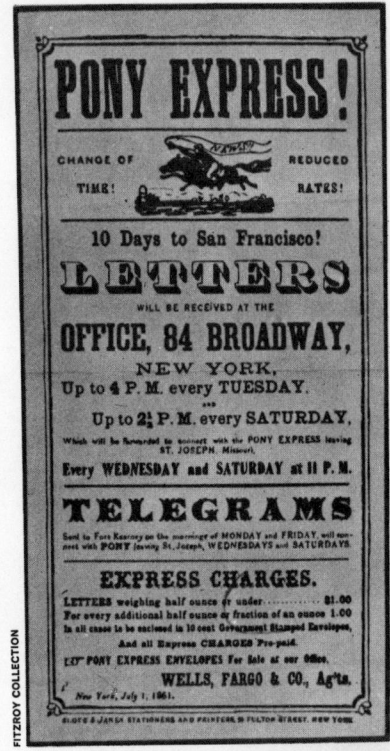

The Western Pony Express, most dramatic of America's early postal systems, was short-lived. It was inaugurated on 3 April 1860 and went summarily out of business in October 1861, when the trans-continental telegraph was completed.

Sailing out over the housetops of Paris, a mail-carrying balloon gives contact with the outside world during the siege of the town by the Germans in the war of 1870. Balloons carried some three million letters, scores of passengers—and hundreds of homing pigeons. The pigeons provided an inward postal service, carrying many thousands of micro-miniaturized messages to the capital from beyond German lines. The pigeons sustained some casualties, but with multiple duplication of microfilms, and as many as ten birds carrying the same batch of messages, most of the 'air-letters' got through.

help was coming from the rest of France—and when.

On 21st September a special instruction was issued from the Director of Posts: all letters for the outside were henceforth to be written on lightweight paper; they were to be folded and sealed. No envelopes were to be used.

On 23rd September a superannuated observation balloon, the *Neptune*, took off from the Place Saint-Pierre in Montmartre, rose to some six thousand feet and sailed majestically over the German lines. It carried a hundred and twenty-five kilograms of mail. Staying aloft till it reached the safety of unoccupied country near Evreux, it came down neatly near a railway station; its pilot took a train and delivered his dispatches personally to the secondary seat of government at Tours. Paris had made contact.

The *Neptune* was the forerunner of many such operations. On 26th September the government of Paris formally announced the setting up of an airmail service. Lightweight letters and postcards soon floated out of Paris by the thousand. Before long, newspapers were being printed on airmail paper. Soon an airmail postal order service was introduced.

Then came something new, the *inward* postal service—mail from the outside world to Paris. In November British newspapers carried an announcement: *Pigeon posts to Paris: Notice has been received at the English Post Office that a special despatch to Paris by means of carrier pigeons has been established at Tours, and that such* despatch may be made use of for letters originating in the United Kingdom . . .

The balloons were now regularly carrying pigeons. Taken from their random landing points all over the countryside to a central launching base at Blois, the birds were liberated for the return journey in batches of five. As an insurance against loss by accident (or by rifle fire) each bird carried a facsimile of the same despatch.

The demand was overwhelming. Messages poured into the despatch office from all over France and from all over the world. Teams of clerks transcribed the messages in tiny handwriting on to lightweight paper to be clipped to the legs of the birds.

Soon it was evident that each pigeon was being under-utilized. France's up-and-coming photographers pointed out the possibilities of miniaturization. With microphotography, a large number of messages could be included on one large sheet of paper and then reduced to a microfilm. On the bird's arrival in Paris the film could either be enlarged as a photographic print, or merely projected as a legible image on a screen for copying by hand. Within a short time postal photographers were packing as many as two thousand five hundred messages into one film. A single pigeon could carry a dozen or more such films at a time; one successful flight meant thirty or forty thousand messages through the lines.

As an exercise in imaginative improvization the scheme has few precedents; in technique it was some three-quarters of a century ahead of its time. It was not until the 1940s, with the wartime 'aerogramme' photoletter service, that the principle appeared again.

The true beginnings of today's communications lie farther back than 1870. The balloons and pigeons of Paris, dramatic as they were in their impact at the time, were in reality a sideshow in the story of communications. A hundred years before when men were tinkering with the possibilities of electricity Louis Odler, a Swiss, writing to a friend in 1773, toyed with the thought of using it to send messages along wires. There was already no doubt that electricity travelled along wire and that it did so very quickly: 'Perhaps you will be interested to learn that I am contemplating experiments which will permit an exchange of views with, say, the Emperor of China, or the English, or the French—or indeed anyone in Europe. Any message could be sent over a distance of hundreds of kilometres in less than half an hour. Does this not merit fame? However these experiments develop, they will certainly lead to great discoveries. But for my part I have not the heart to carry them out this winter.' It was at about the same time that a professor in the University of Padua in Italy, Alessandro Volta, suggested that by the use of wires it might not be impossible for an electrical charge in Como to fire a pistol in Milan . . .

Throughout the century that preceded the siege of Paris men had had an eye on electricity as a messenger. Faster than runners, faster than ponies, faster even than pigeons—it looked a likely starter. In the century that followed it was to lead all the way.

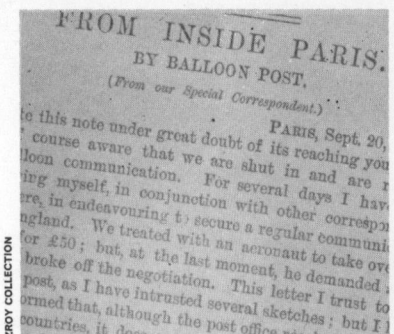

Enterprising press correspondents in Paris during the siege took pride in heading their despatches 'by balloon post'. Sketches and photographs were also sent by balloon. At one stage, foreign newspapers were sending microfilm of their own front pages by return of pigeon.

The semaphore idea goes back to the ancient Greeks. This was the up-to-date version in 1842, signalling. In clear weather it signalled Admiralty and merchant shipping information between London and the south coast.

A WORLD OF INSTANT CONTACT

Electricity Leaps the Frontiers

IN THE PRIVACY of his cabin the captain checked over the printed passenger list. He stopped at the entry for a certain cabin number and read the names: *Mr Robinson and Son*. He was unconvinced. Quietly, he left his own room and made his way to the cabin in question. Letting himself in with a pass key he glanced round quickly. There was no great hurry; he had himself arranged for the Robinsons to be invited to lunch at the captain's table; lunch had just begun, the Robinsons were duly seated, and he had slipped away for a quick look round.

The captain's name was Kendall. The ship was the Canadian Pacific liner *Montrose*, a few hours out from Cherbourg, bound for Canada. The date was 20 July 1910. 'Mr Robinson and Son' were Hawley Harvey Crippen and his disguised secretary Ethel Le Neve.

A brief inspection satisfied Captain Kendall that his suspicions about the pair were well founded. The boy's felt hat, hanging on a hook in the cabin, had paper packed around the rim to make it a tighter fit. Near the washbasin was a face flannel. But it was not really a face flannel; it was a piece of woman's bodice.

A day or so later Mr Robinson and 'son' were on the boatdeck when Captain Kendall passed. The crackle of the ship's wireless transmitter could be heard as he acknowledged them. 'A lovely day,' said Mr Robinson. 'Beautiful,' said the captain.

A few hours earlier Captain Kendall had handed a message to his wireless operator for immediate transmission to Liverpool: *One hundred and thirty miles west of Lizard; have strong suspicions that Crippen London cellar murderer and accomplice are among saloon passengers. Accomplice dressed as boy. Voice, manner and build undoubtedly a girl.*

'A wonderful invention,' said Mr Robinson, nodding in the direction of the sound from the wireless room. 'It certainly is,' said Captain Kendall.

As the *Montrose* steamed on she was slowly overtaken by another, faster, ship, the White Star liner *Laurentic*. The faster liner had left England sometime after Captain Kendall's message had been received. On board were Chief Inspector Dew and Sergeant Mitchell, both of Scotland Yard.

When the *Montrose* arrived at Father Point, at the mouth of the

The 1900s saw the real beginning of 'electric communications'. The telephone (and the 'operator') became a feature of everybody's life. The girl in the picture worked for Britain's National Telephone Company in 1905.

St Lawrence River, she 'stopped to pick up pilots.' Two men went to the captain's cabin and Mr Robinson was sent for.

'Let me introduce you,' said Captain Kendall. Mr Robinson stretched out a hand and it was seized. 'Good morning, Dr Crippen. Do you know me? Inspector Dew from Scotland Yard.'

That same day another message went from the *Montrose;* it said: *Crippen and Le Neve arrested. Wire later. Dew.*

The Crippen case attracted a lot of attention, partly because of the sensational nature of the crime and partly because it was the first recorded instance of a fugitive being brought to justice by wireless. Whatever else Dr Crippen did or did not do, he put telecommunications firmly into the public mind. Electricity, as well as travelling along wires, could now travel through the air; in only forty years the balloons and pigeons of the Siege of Paris had become not merely obsolete but archaic.

Guglielmo Marconi was born just four years after the last pigeon reached Post Office Headquarters in Paris. He was only twenty-one when he sent the first radio message—a transmission without the use of wires. Six years later a 'wire-less message' spanned the Atlantic. Marconi brought in a new era of world communication. But in the meantime there had come into being an intricate network of landline and cable communications. This now linked most of the countries and continents of the world; it was a network which Marconi's 'wire-less' system served to augment rather than to supersede.

The landline had got off to a shaky start. Francis Ronalds, after demonstrating a form of electric telegraphy in 1816, put up the idea to the British Naval authorities as an improvement on the existing signal-arm semaphores. The suggestion was declined: 'Telegraphs of any kind are wholly unnecessary and no other than the one in use will be adopted.' But by the 1840s and 1850s the electric telegraph, much advanced on Ronalds' device, had been universally adopted. It was primarily in the sending of signals between railway stations that its value was first recognised. Its use ran neck and neck with the rise of the railways. Soon, whole countries were served by the electric telegraph. Next came the problem of getting across the sea.

The big talking point in the press of the 1850s was the Submarine Cable. This had a shaky start too; the difficulty was keeping the water out of the wire. The first undersea cable, laid after much trouble between France and England, was finally connected on 28 August 1850. It functioned satisfactorily for as long as it took to transmit telegraph messages to Prince Louis Napoleon Bonaparte—and failed altogether on the following day.

Despite all discouragement, however, submarine cables continued to be laid and re-laid, severed, lost grappled and salvaged, lost and recovered again, reconnected, replaced and improved.

Soon, predictably, eyes turned to the Atlantic. With a telegraph network covering much of Britain and part of the eastern shores of

Samuel F. D. Morse was an artist. (The picture is a self-portrait.) He was also, though less naturally gifted for the job, a dedicated innovator. Faraday's experiments with magnets and sparks inspired him with the idea of inducing sparks at a distance, sending them as a code, by wire. His first transmitting instrument, unnecessarily elaborate, featured movable slugs in a sliding plate. These, pushed along a track, alternately made and broke an electrical contact. The message was 'set' in advance on the plate in the manner of printers' type. Later Morse devised a simplified version of the code and tapped the message by hand.

the New World, something clearly had to fill the gap. Cables had become more rugged and more waterproof; all that was needed, said the cablemen, was cable—1,843 miles of it—preferably in one continuous run. It would take a year or two to lay, but when it was done it would be magnificent.

It took thirteen years. Again, there were breakages, splicings, electrical faults, losses and recoveries. Again, but at infinitely greater cost, there were transmission failures. Again there were apparent successes, congratulatory messages, celebration banquets and honours—and aftermaths of total silence.

The cable-laying vessels, British and American ships recruited for the job by private enterprise, performed prodigies. The two ships struggled, each with a hold-full of half an Atlantic's length of cable, first to lay the cable from mid-way to opposite shores, then to take it in turns from west to east. There were two major attempts: the first, in 1857, ended with a broken cable two thousand fathoms deep in mid-Atlantic. The second, in 1858, involved two separate expeditions; the final connection, completed in August 1858, failed after operating for three months. Seven years later the reconditioned *Great Eastern* was brought in. By 7 September 1866, after disappointments and disasters to match her scale, Brunel's big ship had succeeded in laying two operational cables from Valentia, in Ireland, to Trinity Bay, Newfoundland.

Cyrus W. Field, the American magnate to whose enterprise the project was indebted, did not conceal his relief at its completion. His ship-board journal conveys the strain that twelve years of endeavour had laid on him. 'Never shall I forget that eventful moment when, in answer to our question to Valentia, in an instant came back the memorable letters *O K* : I left the room, went to my

When street telephone boxes were installed in Chicago in 1893 a magazine artist visualized their usefulness in an imaginative pre-view; a policeman calls for aid after a street accident. The drawing presented an archetype of countless thousands of such situations to come. In the 1970s picture above, a Swiss railwayman calls for aid after a landslide. It must be said that the 1890s had few scientists who failed to see the potential of electricity in communications.

As well as being undeniably useful, the telephone rapidly became beautiful. This was a 'local battery hand combination' instrument in Britain at the turn of the century.

Before telephone and telegraph wires went underground, big cities grew forests of multiple poles, with skylines fretted with many hundreds of lines. This was Broadway, New York, in about 1880. On the left, a focal point of the system, is the Western Union building. In 1887 New York set up a Board of Electrical Control. The organization had the daunting task of placing all of the city's electrical wiring underground. With a momentary lapse of foresight the Board reported that it could 'hardly anticipate success in having all wires placed underground, because in some parts of the city the demand for electrical service is so scant that it would not pay to place expensive conduits in such districts'. When, in 1889, the Metropolitan Telephone Company's New Central Station and Great Switchboard was opened there was little doubt as to its ability to cope with demand. 'It is presumption to set a limit on an invention,' said the 'Scientific American', 'but the multiple switchboard seems to have nearly reached perfection . . . the connections are so arranged that any operator, without leaving her place, can connect with any subscriber . . .' There were at the time two thousand five hundred subscribers but the switchboard had capacity for up to 'as many as ten thousand.'

1864: the Prince of Wales pays a ceremonial visit to the cable tank of Brunel's converted 'Great Eastern'. The ship was the only one in existence capable of carrying the full two thousand miles of cable needed to span the Atlantic. The Prince's message of good wishes was passed through the length of the cable as it lay coiled in the ship; it travelled instantaneously over the two thousand miles, finishing where it started. The Prince's good wishes were in vain, however. Success was to come only on a second voyage two years later.

cabin and locked the door. I could no longer restrain my tears.'

For all their air of confidence, the innovators of the 1860s and 1870s were continually amazed at their own success. But amazement and success had become collective. By this time each new prospective step in communications development had become apparent not just to one mind but to many: electricity was running away with itself. Alexander Bell, whose sole authorship of the telephone was even at the time open to question, said, 'I must admit I am amazed it should be possible for someone to speak in Washington and be heard by someone else at the foot of the Eiffel Tower. I may have pointed out the way, but many other inventions and the collaboration of many minds made the achievement possible.' Like Morse before him—whose coded telegraph message between Washington and Baltimore in 1844 had read: *What hath God wrought?*—Bell saw himself less as inventor than

discoverer. The transmission of actual speech along a wire, as opposed to simple coded impulses, was thought of almost literally as a miracle.

Before long the miracle settled down to common acceptance. By the 1880s, with the addition first of manual 'exchanges' in which users could be put in touch with each other merely by stating a number to an intermediary, and, soon after, with the advent of automatic exchanges, the telephone was everbody's gadget. Even Queen Victoria and the Pope had one.

In some cities, in addition to using the telephone for business and social purposes, subscribers could be connected to the 'electrophone.' This, for a moderate yearly fee, allowed the citizen to eavesdrop on theatres and concert halls. The service in London offered two tariffs: the cheaper rate covered the use of two receivers, and the *de-luxe* version not only doubled this number but, as the telephone directory stated, allowed the subscriber 'the right to select performances to be heard'.

The coming of the electric age brought changes right across the world. The transmission of information, geared in the beginning to the tempo of the running man and the trotting horse, then to that of the locomotive and the steamship, now became instantaneous.

But still there remained the need to transmit physical objects—the letters, parcels and packages that could not conveniently be converted into electrical impulses and radio waves. All over the world ordinary people were waking up to each other's existence; the mails, which for centuries had remained the instruments of monarchy and government, at last came within the reach of everyone. It was Rowland Hill, British schoolmaster and civil servant, who achieved this.

Prior to 1840, when the British government adopted his proposals for a uniform penny post, postage had been charged on letters according to the distance they travelled. This meant, at rates which were high even for short distances, that longer distance post was more or less out of the question for the ordinary correspondent. It also meant expenditure of time and trouble in working out the cost of each letter. The system was open to abuses: postage was generally collected at the end of the journey—from the recipient rather than the sender—and the recipient had the right to refuse both the letter and the charge. Postal authorities were plagued not only with accumulations of undelivered letters, but with handwriting codes adopted by people who conveyed messages to their friends on envelopes free of charge. Once the proffered envelope had been seen, and its message understood, the 'letter' was declined. A further abuse was the forging of the signatures of those distinguished people whose mail travelled without payment; the flourish of a famous name as sender on an envelope was easily imitated.

Rowland Hill, a methodical and logical man, saw that all these snags could be done away with. By simple mathematics he

The new jungle telegraph. The picture shows Royal Engineers setting up 'electric telegraph wires' along the line to Prah-su during the Ashanti War of 1872-74.

MARCONI

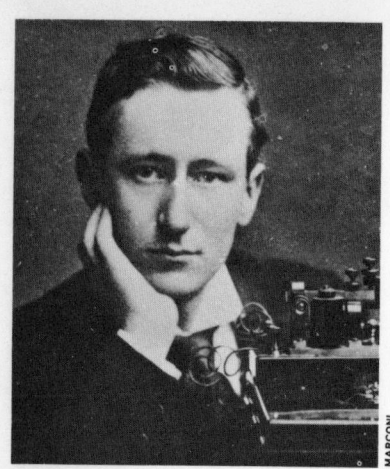

At the age of twenty Guglielmo Marconi, a student in Bologna University, read an article in a magazine about the work of Heinrich Hertz on electro-magnetic waves. Stimulated by what he read, he set to work on improving methods of transmitting electrical singals. Within a few months he had succeeded in transmitting morse through the air. In less than a year he discovered the principles of 'earthing'. In 1896 he approached first Campbell-Swinton in London, then, as the letter indicates, Mr Preece of the General Post Office. Within two months he had filed a patent for 'improvements in transmitting electrical impulses and signals and in apparatus therefor'. Less than a year later the Marconi Wireless Telegraph Company was formed.

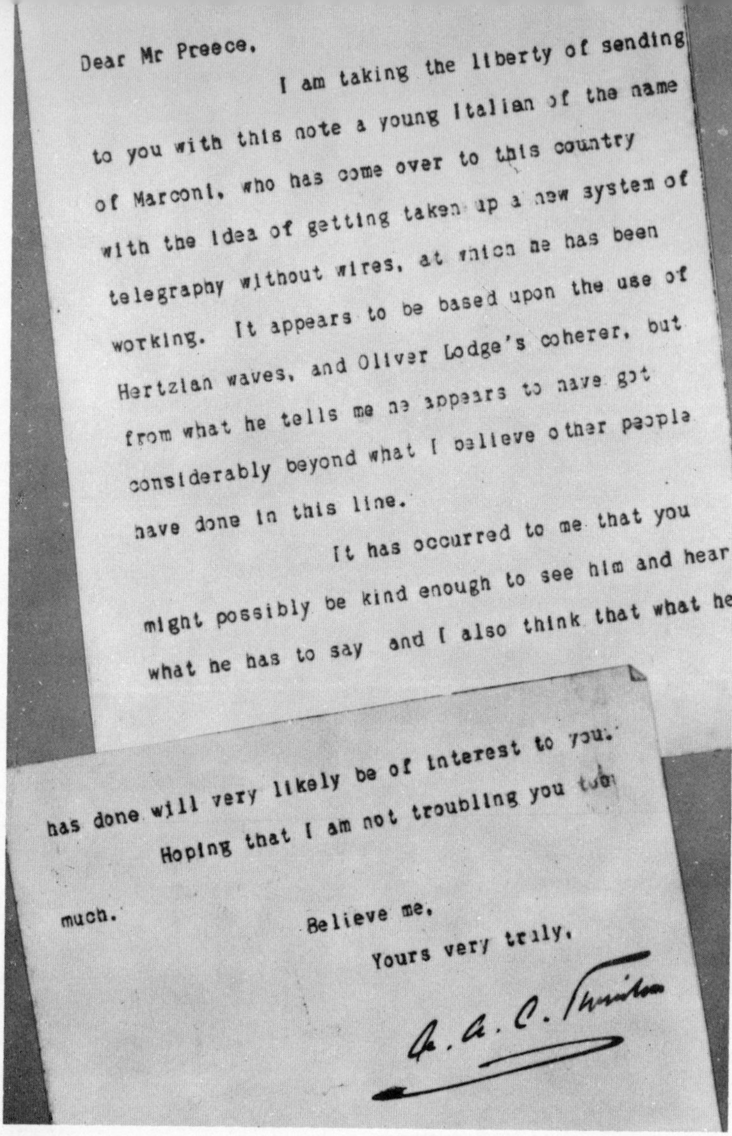

Dear Mr Preece,

I am taking the liberty of sending to you with this note a young Italian of the name of Marconi, who has come over to this country with the idea of getting taken up a new system of telegraphy without wires, at which he has been working. It appears to be based upon the use of Hertzian waves, and Oliver Lodge's coherer, but from what he tells me he appears to have got considerably beyond what I believe other people have done in this line.

It has occurred to me that you might possibly be kind enough to see him and hear what he has to say and I also think that what he has done will very likely be of interest to you.

Hoping that I am not troubling you too much.

Believe me,

Yours very truly,

A. A. C. Swinton

MARCONI

showed that, with postage paid by the sender, one penny as a general rate would make the Post Office a going concern. He showed that it cost less to carry a large number of letters a long distance than a small number a short distance. He proposed, in a suggestion that was to have repercussions all over the world, that the Government should sell small receipts for prepayment of postage—'a bit of paper just large enough to bear the stamp'—to stick on the envelope to show that postage had been paid.

Rowland Hill's scheme was adopted. It worked. In a few years the postal system had thrown off its function of tax collector and almost, but not quite, forsaken its role as a communications medium for Authority. From being an instrument of imposition it had become a social service.

By 1845 the United States had followed Britain's example. Canada and France were next. Adhesive postage stamps were introduced in one country after another. Prepaid covers appeared;

by 1870 thirty countries were issuing stamped envelopes.

In 1840, the year Rowland Hill's scheme was introduced, the number of letters posted in Britain soared from eighty million to nearly one hundred and seventy million. By 1870 the figure had reached some nine hundred million. Throughout the world, postal traffic boomed; as the network of telecommunications grew, so—less spectacularly but as significantly—did the network of mails.

The obvious became rapidly clearer: communications, by their very nature, are of more than regional concern. Soon it was seen that national networks must inter-operate; soon there had to be international agreements on methods, standards and costs. Who was to pay the cost of sending a letter from Montreal to Los Angeles—the Canadians, over whose territory it travelled only seventy miles—or the Americans, who had to carry it 2,500 miles? Who was to pay whom for a telegram sent from country A to country B over the cables of countries C D and E?

These and many hundreds of allied problems had to be settled. In 1863, on the initiative of Montgomery Blair, Postmaster General of the United States of America, a postal conference was called in Paris, it was the first step in the discussions which were to lead to the signing of a 'Treaty Concerning the Establishment of a General Postal Union'. In 1875 a convention was signed. In 1878 the General Postal Union became the Universal Postal Union. Today the Union's member countries, over one hundred and thirty of them, cover almost the entire population of the world.

Behind a prosaic title lies a concept brilliant in its simplicity and masterly in its breadth of vision. Member countries are considered as forming 'a single postal territory for the reciprocal exchange of letter-post items'. Stemming from this concept is a secondary proposition: that member countries give freedom of transit to mails passing across their territory from other member countries. The implications of these twin propositions—alone of the many Articles of the Conventions—are far-reaching. They represent a degree of cooperation almost unparalleled in human history.

But there is a parallel. In 1865, in a completely independent operation, the telegraph authorities of the world similarly decided to talk over problems of internationality in telecommunications. At first concerned only with the electric telegraph, the organization that emerged was later to widen its scope to embrace the whole field of telecommunications. Today the International Telecommunication Union is the single international authority governing the operation of telegraphy, telephony and radio around the world. It was born in the same epoch as the Universal Postal Union, and for the same reason.

The two organizations, both now specialized agencies of the United Nations, epitomize communications in the larger sense. For the 1870s they represented a point of culmination. For the 1970s, enlarged and consolidated, with the respect of the whole world solidly behind them, they represent a point of new departure.

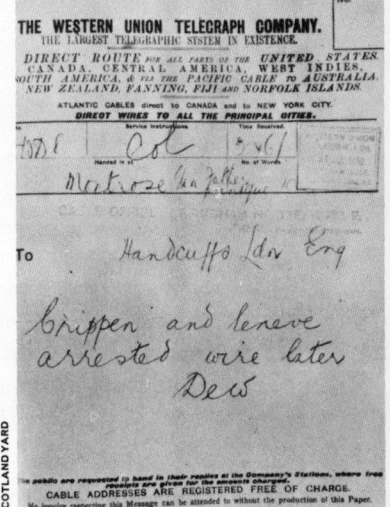

This was the wireless message received at Scotland Yard after Inspector Dew had arrested Dr Crippen and his secretary on their arrival in Canada. It was the last of an historic transatlantic exchange. Captain Kendall had signalled his suspicions to London by wireless. Back came a reply from Inspector Dew, who was by now on board a faster ship and ready to intercept the 'Montrose': 'Will board you at Father Point'. Captain Kendall replied, 'Shall arrive Father Point about 6 am tomorrow. Should advise you to come off in small boat with pilot, dressed as pilot.' The arrest was the first ever to be effected with the aid of transatlantic wireless. The public, till then hazy about the value of the new invention, now woke up to it. Ethel Le Neve was acquitted. Crippen was hanged.

POST-BOX OR HAND-SET?

The Electronic Way from A to B

THE COMMUNICATIONS INDUSTRY is concerned with two distinct forms of material. On the one hand there are items that can be translated into electrical impulses for transmission and translated back into their original form on arrival. Television pictures, telephone speech and teleprinter characters are typical of the translation principle; moving images, speech and letters of the alphabet are first changed into electricity and then reconstituted. The process—once the technical problem of transmission and re-translation is solved—is relatively easy.

On the other hand there are items that are less readily translated, or which people prefer to receive in their original form. Such items as contracts for signature, parcels, packets, and perfumed love letters have to be carried from one place to another just as they are. No matter how ingenious the handling and routing solutions, this process is distinctly less convenient.

The number of physical objects requiring carriage from one place to another over a year has become astronomical. At the time of the wave of postal reform in the 1840s and 1850s, people were amazed at the published postal statistics: in Britain, to take just one example, weekly turnover of postal items was over one and a half million. After the introduction of the penny post the figure rose higher. In three years it had reached four million. By the 1960s the weekly total was around two hundred million, and the United States figure was one thousand two hundred million.

With the growth of literacy, and with the growth of populations, the mails have grown to a tidal wave—a daily tidal wave, growing each day. By the early 1950s the experts saw that the upward curve was dangerously steep. It was clear before very long that something radical had to be done if the world's postal services were not to be swamped. The old techniques of postal sorting, handling and transportation—many of them hardly changed since the days of Franklin and Rowland Hill—were overdue for reassessment Also due for an overhaul was the whole concept of postal delivery; could the standards set by previous generations be maintained?

In more leisured and less crowded days, pioneers of the mails had striven for a simple objective: universal person-to-person

Advanced microwave transmission techniques need power. The picture shows some of the aluminium feeders (red, positive; blue, negative) in the power room of London's six hundred and twenty foot Post Office tower.

In a seemingly unending flow, parcels are conveyor-belted along the upper ledge of the parcel slope or 'glacis', at London's big Mount Pleasant sorting office. To even out the distribution of the parcels on the slope, a remote-controlled diverter moves from point to point along the ledge as required. Men standing at the foot of the glacis pick parcels from the slide and place them in appropriate sorting bins.

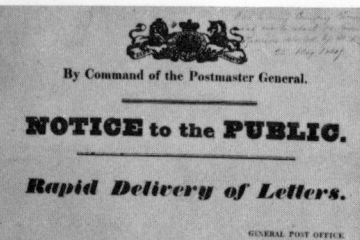

In 1849 the full significance of universal door-to-door delivery was beginning to be realized. This appeal from the Postmaster General, asking householders to help by providing 'street door letterboxes or slits' was symptomatic. Though expressed in the politest terms, the request was first of many—and ever more anguished—cries from the postman's heart.

contact throughout the world. There were some areas where distance and inaccessibility made front-door delivery impossible; even in the 1970s some places have to settle for delivery at a local collection point. But for the most part, the service has become 'all-the-way-home'—post-box to doormat. In countries like Britain, Holland and Denmark, where the postal territory is small, the objective has been even more demanding—not only post-box to doormat, but *guaranteed delivery anywhere in the country by first post the following morning.*

To the man in the street the proposition seemed scarcely practical. Delivery to any one of twenty million addresses by first thing the next morning? Impossible. As the years passed and the postal figures rose higher and higher it began to look impossible to the experts too.

In a number of countries, there was not only too much mail, but the railways, on whose development the expansion of the mails had greatly relied, began to decline. In the same way that the coming of the railway had swept away the pony express and the mailcoach, big developments in road and air transport now began to push aside the railways. In Britain nationalized railways faced crises of profitability; in America they actually began to go out of business. United States Railway Post Office routes—services given over to the sorting of mails in transit—had in 1925 numbered fifteen hundred. By 1968 the figure had shrunk to forty-six.

As in most other fields in the 1960s new technologies, new social trends, had brought new problems. The world's postal services were reaching a crisis. The old methods of postal handling were worn out; what was to replace them?

The experts looked again at the classic pattern of postal sorting: collection, segregation, facing, primary sorting, secondary sorting and the tenuous thread that brought each individual letter to its

proper destination. It was a pattern of huge complexity.

The work of a big sorting office sets the scene. At reception bays, vans back in with their loads of mail. Scarcely has one van discharged its burden than another moves into place; they bring mail from street post-box collections, bulk mail from business houses, mails in transit from other centres—an apparently unending flow.

As the contents of the bags spew out on the processing tables inside the building, the first big job begins. Out of the deluge of letters and packets of every shape and size a little order is made: working swiftly, shoulder to shoulder, postal staff stand at the tables removing awkward items by hand. These, the oversize items, the thicker envelopes and packages, cannot be fed through the equipment that stands ready to process the mail; they are dealt with separately (and more slowly) by hand.

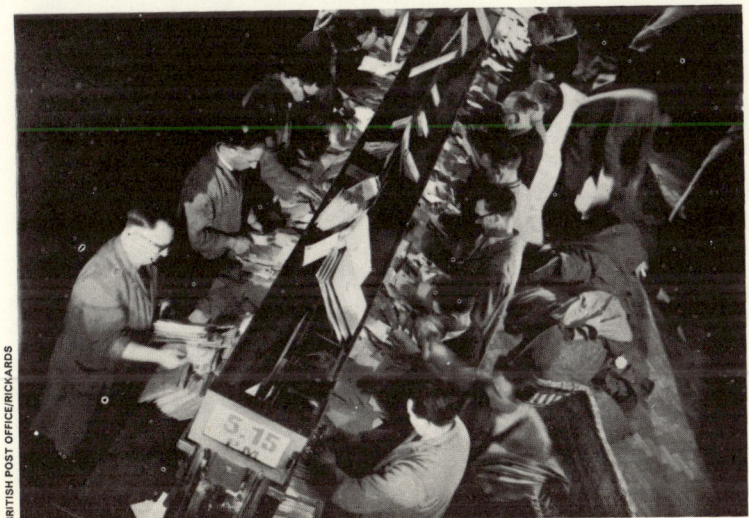

'Facing' letters is an operation which has not basically changed for over a hundred years. A hand-operation, it involves arranging the mail in manageable order—first removing odd-shaped items for special treatment and then placing all of the rest face up, right way up. With all stamps uniformly positioned letters can then be fed at high speed into automatic cancelling machines. New techniques now being adopted by most postal authorities allow for automatic facing.

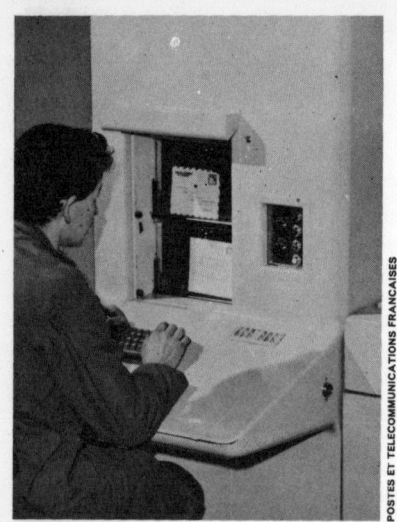

Automatic facing and sorting of letters uses the 'magic eye' principle. First, reacting to invisible fluorescence in the stamps, a machine tilts and manipulates the envelope until it is correctly positioned. Then each envelope is automatically presented to an operator at a keyboard (above); he reads off the written postal code and, by pressing appropriate keys, invisibly prints the code on the envelope. From here the coded envelope is fed into a high-speed sorter (below). This 'reads' the code and routes the item to the next point on its postal journey.

Next come the facing tables. Here the mass of letters is brought further under control. Again with rows of men standing at the tables, all the envelopes and postcards are arranged by hand in manageable piles with stamps facing upwards, each in the bottom left-hand corner. The orderly piles of 'faced' mail are now moved to the cancelling machines. Zipping through them at some six hundred a minute, the machines count and cancel them in one operation. The letters are fed into the machine by one operator and moved by another; they go through so fast that all the eye sees is a blurred white streak.

The letters are now ready for 'primary' sorting. This operation is to break the mass of mail down in geographic areas. Again it is a hand operation. Seated at batteries of pigeonholes, sorters distribute the faced and cancelled letters into appropriate named compartments, reading each address in turn. Each battery of pigeonholes is a duplicate of its neighbour. By multiple duplication the entire mass of letters is handled in the same way, the gigantic process of address reading and pigeonholing being portioned out among the men.

But this is only the first sorting—a division of the mass into basic geographic areas. Now comes a further winnowing. The bundles of letters are removed from their compartments and conveyed to secondary sorting points for breaking down into specific towns. Again, it is handwork at batteries of pigeonholes.

Now finally placed in bags bearing town-name tags, the letters are joined by the awkward items that left the stream for hand attention at the beginning. The progress of these stragglers is noticeably slower. Many get left behind for the next wave. The completed bags are ready for the big move—as a rule, by train.

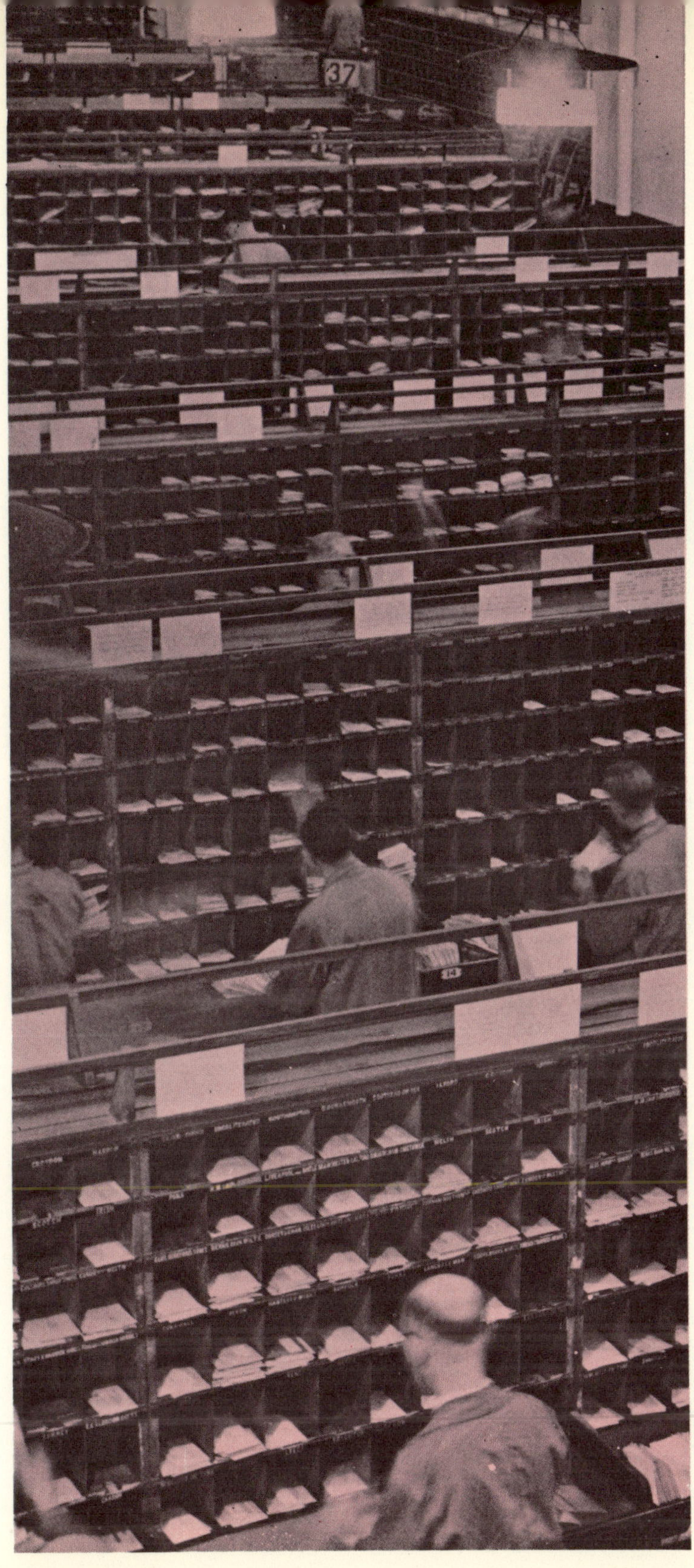

BRITISH POST OFFICE/RICKARDS

BRITISH POST OFFICE/RICKARDS

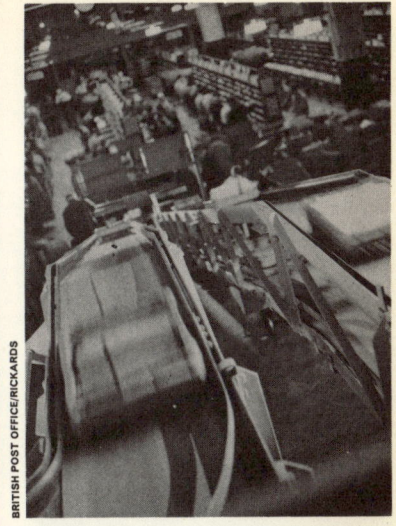

The picture on the left gives only a limited impression of the scale of the main letter-sorting office at Mount Pleasant, London. The section covers over three acres. More than three thousand men are employed here on a round-the-clock shift basis, each dealing with a forty-eight pigeon-hole sorting unit. In all there are over fifty-five thousand pigeon-holes as well as nearly two thousand hanging mail bags. The office deals with an intake of fifteen thousand bags of mail daily. The smaller picture shows part of the overhead conveyor system. Baskets of sorted letters are individually routed round the building for automatic discharge at dispatch points.

In Britain, and in many countries in Europe, the Night Mail is the backbone of the postal service. The trains, unlisted on public timetables, and many of them ten or a dozen coaches long, are given over entirely to the mail. As well as carrying it, they sort it as they go. While some of the coaches are used only for stowage, others are fitted up as travelling sorting offices. Here, as in the conventional sorting office, men continue to subdivide the contents of the mailbags, pigeonholing items according to destination. Working to carefully timed schedules, they arrange their work to allow take-up and despatch of mail at points along the line.

For most of the transfers the train does not even slacken speed. Trackside apparatus snatches mail from an extended arm as the train passes at speed. A similar system snatches mail from prepared collecting points; transfer pouches are scooped into the train at eighty miles an hour. With almost explosive impact they thud into the body of the collecting coach—sometimes three or four in succession.

In a short time the delivered bags are being driven away from their trackside arrival points in postal trucks; within the travelling sorting office, collected bags are opened and sorting begins for the next delivery point. The process, vital to the concept of overnight-delivery-anywhere, is a remarkable exercise in high-speed transfer. Over the years it has become a routine as unremarkable as the rest of the postal service. (It must be said that it has had plenty of time to settle down. The first such trackside snatch was effected in 1842 and, although train speeds have somewhat increased, the principle has remained virtually unchanged since then.)

As they near their destinations, letters again require to be sorted —sometimes, depending on the complexity or otherwise of the route they have to follow, even twice more. The final sorting, where the system provides for delivery to the recipient's front door, is the postman's. Before setting out on his round he arranges his mail in sequence. From personal experience he knows the time-and-leg-saving-route for his particular 'walk'.

Every day of every year the world's postal authorities move mountains of mail. Most of them, as well as handling letters, packets and postcards, also accept parcels. These, even less manageable than ordinary mail and certainly less susceptible of automatic processing, are also on the increase. Their weight, volume and lack of uniformity are a growing problem.

London's Mount Pleasant sorting office handles four million letters and one hundred thousand parcels a day. Two thousand van-loads of mail move in and out each day and over six hundred loads arrive and leave daily on the Post Office's underground railway.

It is significant that, of the three thousand men employed at Mount Pleasant, some fifteen hundred are occupied in handling a relatively small minority of its individual items—parcels. But both

Technical progress in the handling of parcels has been rapid. This mechanized parcel-sorting hall in Braunschweig, Germany, epitomizes the trend. But parcels traffic demand remains everywhere at a very high level. Germany's annual turnover is in excess of two hundred and fifty million; Britain's figure is some two hundred million; America's —nearly one thousand million.

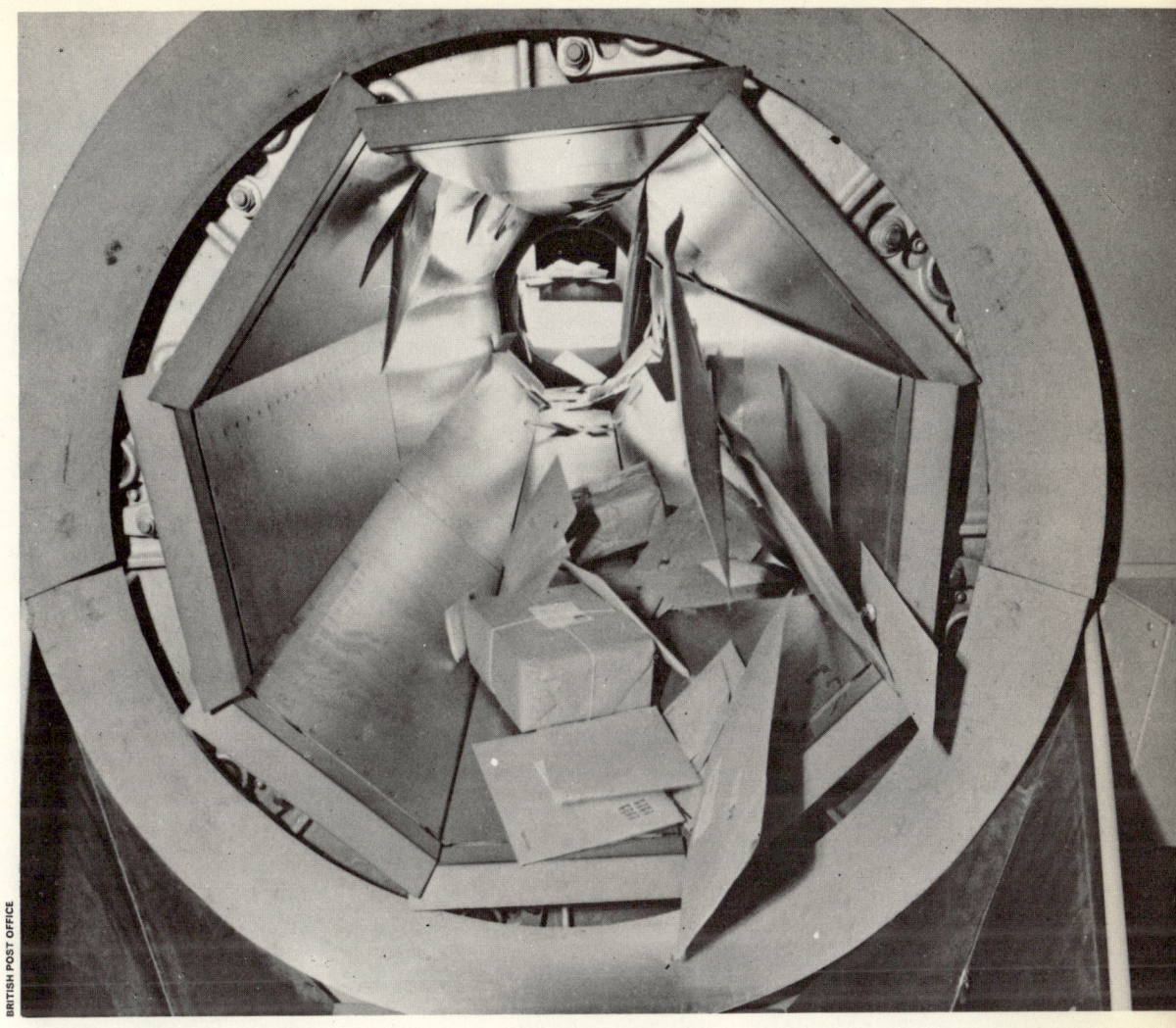

classes of mail, letters and parcels, are becoming increasingly costly to handle.

Soon, somewhere, something has got to 'give'. Postal rates have risen. Deliveries, in some countries initially as frequent as three or even four times a day, have, in most cases, been reduced to two. In the United States it is one. Annual deficits for the postal services everywhere have soared.

Mechanization and automation, once looked upon as an acceptable luxury, now appear as the only way out. Two conclusions emerge. One is that physical transportation of actual objects must, as far as possible, be mechanized; the other, even more to the point, is that physical transportation of actual objects must be discouraged. For the Post Office of tomorrow the only answer is electronic communication, with speech and document transmission carried out instantaneously through wires or through the comparatively uncrowded air.

This machine automatically filters out awkward-sized mail from the mass of postal items. As the inclined drum revolves, slits along its length allow normal mail to slip through to high-speed conveyors below. Awkward items, which cannot be dealt with automatically, are fed out of the open end of the cylinder for hand-processing.

Seventy feet below the streets of London the Post Office runs its own railway. Entirely automatic, its trains operate without drivers or guards. Between stations they travel at speeds of thirty-five miles an hour (noticeably faster than most of the traffic on the surface) and at station approaches a system of inclines serves to reduce speeds automatically to seven miles an hour. An average train carries six-wheeled containers, each carrying fifteen letter-bags or six parcel bags. Containers are wheeled on and off the trains at stations; rotary tippers are used to tilt the containers to discharge loads on to conveyor belts and hoists. Mail is processed in sorting and terminal offices in buildings above the stations. The system has six-and-a-half miles of tunnels, eight stations and some forty trains. On an average day it carries over forty thousand bags of mail. The time saved by the railway, as compared with the tortuous progress of a mail van through London's traffic, varies with conditions in the streets. On a normal single journey between terminal stations it amounts to eighteen minutes. In rush hours, and in bad weather, the saving is much greater. The system is a testimony to the far-sightedness of those who planned and built it; it has required little or no modification since its inception in 1927.

The cable industry, for long identified with the technologies of the nineteenth century, is a spirited growth point of the 1970s. The 'Marcel Bayard' (above) is the latest of France's cable layers. Her four holds have a combined capacity of one thousand nautical miles of heavy-duty cable. The ship has a crew of ninety-three officers and men, a team of twenty cable technicians, and can remain at sea for periods of up to fifty days.

A deep-sea cable is paid out over the sheaves of a British cable-layer. The picture on the facing page shows stored submarine 'repeaters'. These are designed to boost the signal from stage to stage on its way between terminals. They are inserted into the cable at intervals and are 'launched' either through secondary paying-out equipment or, where they occur frequently, by an adjustable machine which accommodates their extra thickness and rigidity without damage.

Postal experts have already gone a long way to implementing both these conclusions. Mechanization of postal handling has been speeded up. Development in speech and image transmission has been spectacular.

First of the developments in mechanical handling has been automatic segregation. Here, to replace the tedious picking and winnowing of the old-style segregation table, letters and packets pass on a conveyor into the mouth of a revolving drum. Inside the

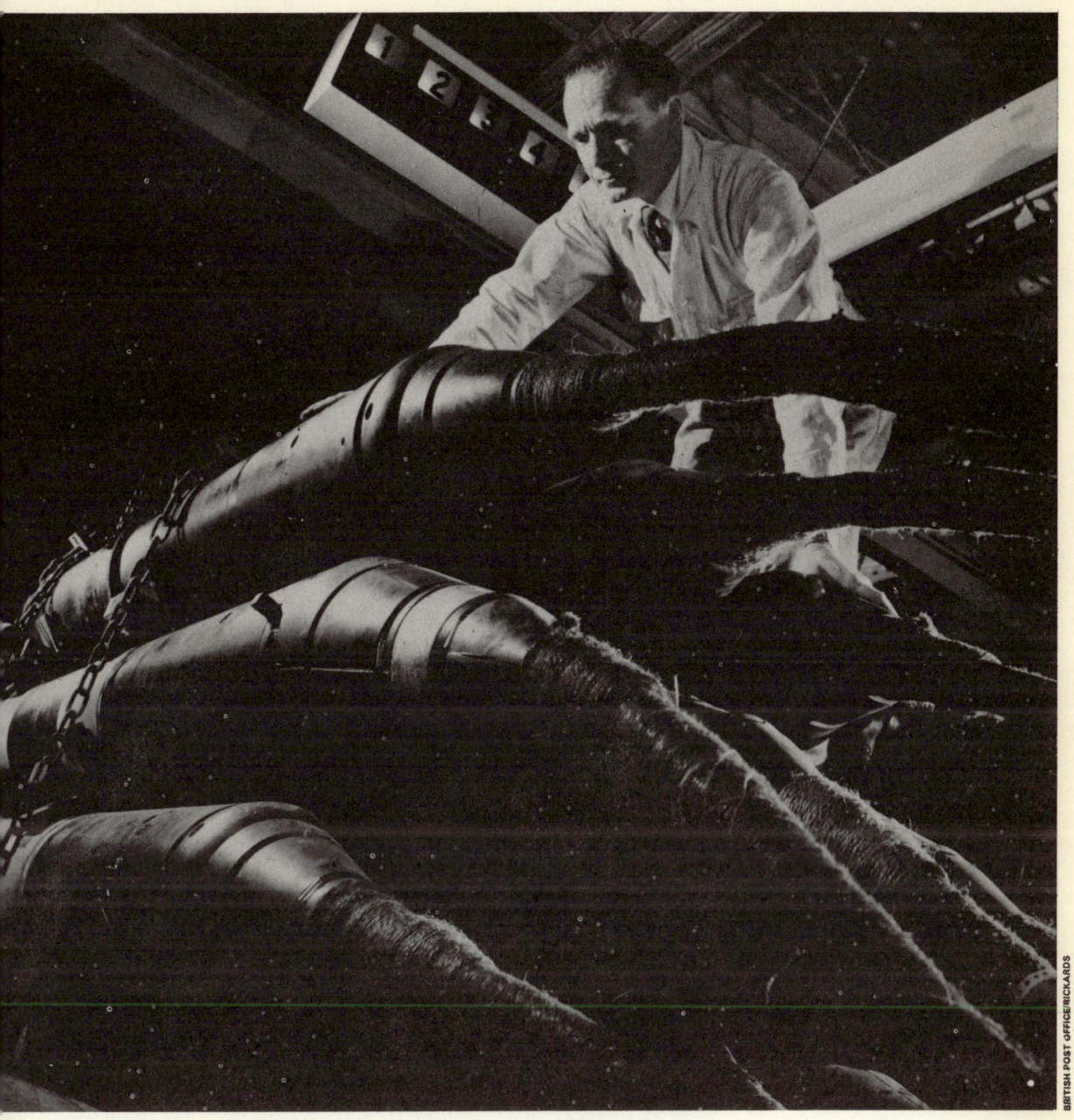

drum, which is tilted towards a further conveyor, the items topple and tip as they are caught in slits in the side of the drum. The thin items slip through the slits to conveyors below; these are the 'easy' items—the majority which can now be mechanically processed. The 'awkward' items fall out at the lower end of the drum for the old-fashioned manual treatment.

Next, for the bulk of the material, comes the automatic facer. Instead of being hand-processed at the facing table, the letters are

Communications are today concerned less and less with movement of physical objects from one place to another, more and more with conversion into transmissible impulses, and re-conversion at the receiving end. This 'handwriting reproducer'—an old idea now miniaturized—transmits the movement of the pen to the distant point.

fed into high-speed facer/cancelling machines. Electronic-eye units examine each item, automatically placing them with addresses upwards and stamps in the right corner for automatic cancelling. In addition, where the postal authority operates a first- and second-class mail service, lower paid mail is separated by the facer/canceller; electronic eyes examine the stamp colours for values, speeding first-class items and, if pressure demands, delaying second-class for handling in the next mail-wave.

These machines are transforming mail handling speed. But the big breakthrough is the postal code. This, used in conjunction with automatic segregation, facing and cancelling, is the key to the help that the postal services need.

The postal code has a number of variants in different parts of the world, but basically the system translates addresses into numbers and letters. Imprinted as phosphorescent dots on the envelope, these allow automatic 'reading' of the envelope at any stage of its progress. The dots are invisible to the human eye but the electronic eye of the sorting machine recognizes the detailed information that they convey, and channels the envelope accordingly.

The problem is to get the phosphorescent dots on to the envelope in the first place. The solution here lies with the public. Every

This was Telegraph House, Trinity Bay, Newfoundland, North American terminal of the first transatlantic cable, successfully completed in August, 1858. Transmissions failed after only three months. Total length of undersea cables today exceeds sixty-five thousand miles—some links carrying eighteen hundred conversations at a time.

address in the postal territory falls within the boundaries of a given code; each correspondent is asked to use it—both on his own notepaper, for use by others, and on correspondence he sends out himself. Added on envelopes to ordinary plain-language addresses, which machines cannot read, the postal code provides the information that they can read.

But between the hand- or type-written characters on the envelope and presentation to the code-reading machine, there has to be a process of translation—a process already noted as basic to the new communications industry. In this instance the information (the coded address on the envelope) is rendered into its phosphorescent equivalent by means of a keyboard. Envelopes are displayed to the human operator in succession; he taps out each code as he sees it; the phosphorescent dots are added to the envelope.

In Britain, where the postal coding system operates right through to final delivery, after the first reading the envelopes need not be read by human eye again until they reach the delivery man himself—the postman who collates them for his walk.

From the coding machine the letters pass to automatic sorters. Various machines are in use; some models sort at slower speeds to a large number of divisions, others sort more quickly to fewer

The telephone, initially conceived for speech alone, now acquires new dimensions. This desk-top instrument allows the ordinary telephone hand-set to be used for the transmission of pictures over the public telephone line. The distant number is dialled; when it answers the hand-set is placed in the cavity and the document, wrapped on the cylinder, rotates under a scanner. A similar machine at the distant number reproduces the image exactly. Transmission time for a document or picture measuring thirty-two square inches is four minutes.

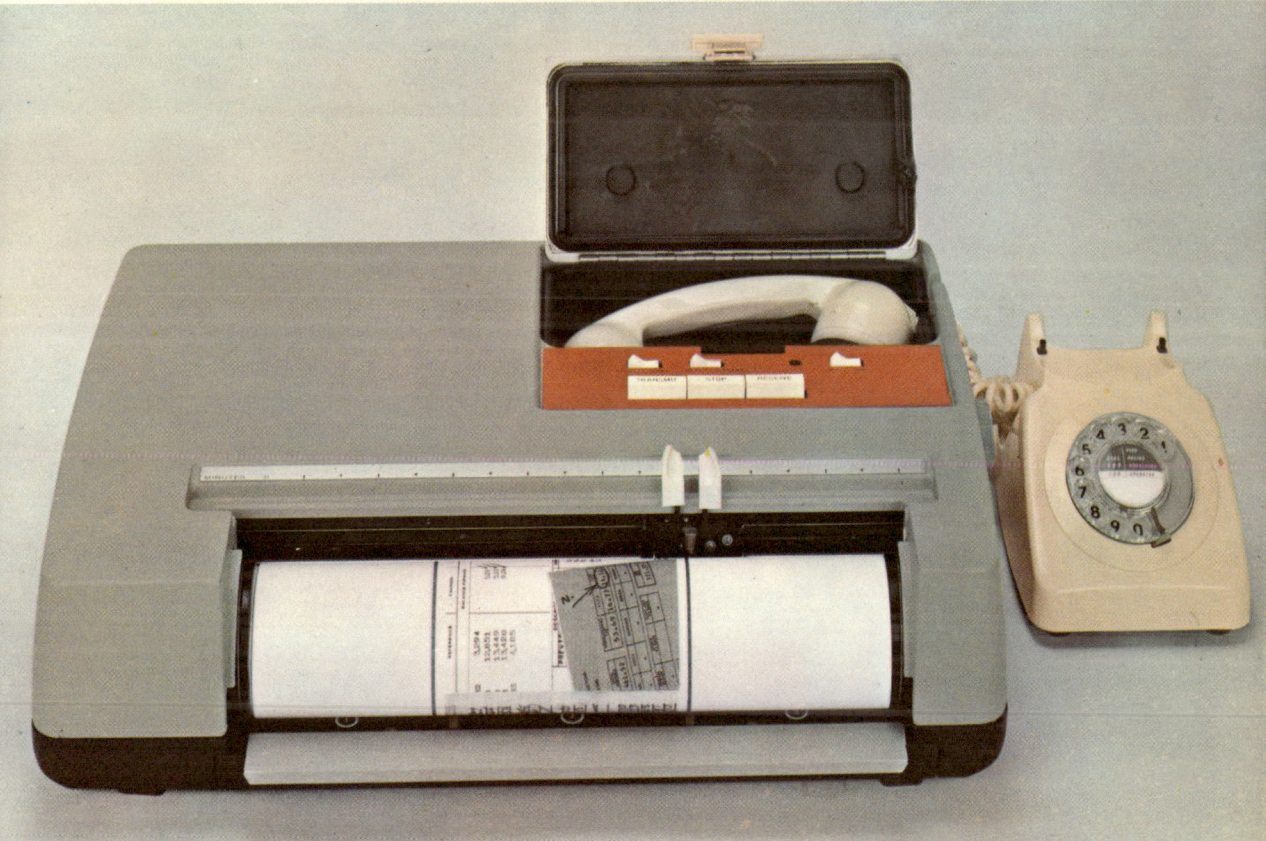

The picture telephone is in course of realization in a number of countries. These pictures show experimental Japanese instruments, one requiring the use of a hand microphone, the other leaving both hands free.

divisions. Sorting to twenty divisions, one machine handles as many as twenty thousand items an hour.

Though it has been a long time on the way, it is clear that the technological revolution in the postal service has arrived; the days of the segregation tables, the facing tables, and the pigeonhole-

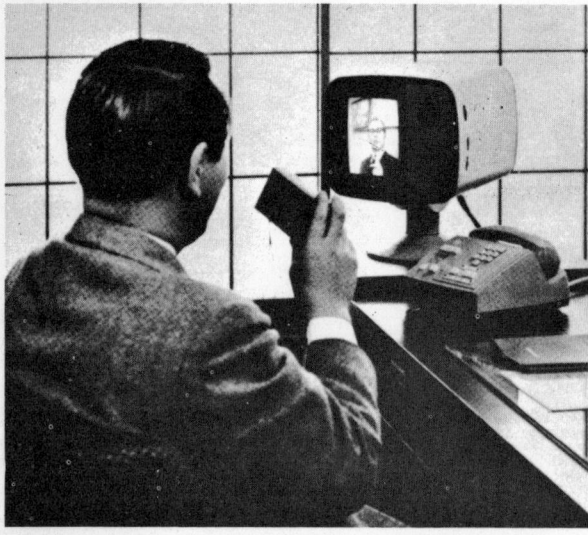

This instrument can be used for private conversation as a conventional handheld telephone, or, standing upright on a desk, it serves as a loud speaking office intercom. Calls can be answered, if desired, from anywhere in the room. 'Dialling' is by push-button keyboard.

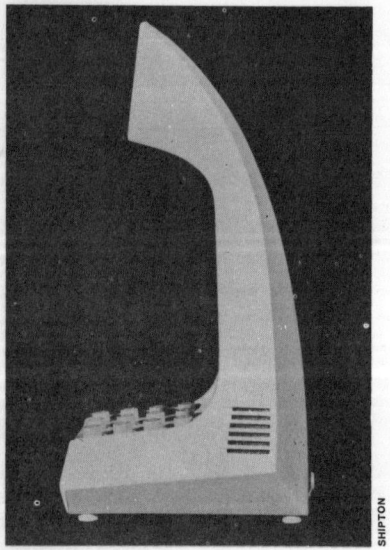

fillers are numbered. Their disappearance now depends not on technology but on full public understanding and acceptance of the code principle. (It must be said that history does not offer much encouragement. Although without benefit of electronics, the postmaster of Chicago once offered correspondents an innovation through a system of 'carrier' numbers. These, added to the normal address on envelopes, were designed to speed mail by channeling it direct to specific delivery stations. The year was 1895. The system failed to arouse much interest.)

On the parcels front, mechanization makes less noticeable progress. Basically the problem is one of weight, bulk and inconsistency of shape and other characteristics. But chutes, conveyor belts, electric-eye control and overhead chain-ways have taken much of the manhandling out of the job. In most of the world's big parcel sorting offices the work is dominated by a large *glacis*, a controlled 'landslide' of parcels, fed from a continuous belt at its upper edge and cleared at the foot by men who heave each item into an appropriate wheeled container bearing a district name.

Here, it would seem, as with letter-post, part of the future lies in standardization of packaging. 'Recommended' envelope sizes will inevitably become *required* sizes. Packages too must come shortly under control. Today's miscellany cannot continue for many years more.

The question must also soon arise as to whether parcels and packets belong to the field of communications at all; today, as in the case of ordinary letter mail, postal experts are wondering how wise they are to consolidate a role which portrays them as round-the-clock hauliers and removal men.

While development in the transmission of physical objects has been slow, in the category of 'translatable' information—speech, numbers and images—it has forged ahead. The problem has been not so much to devise new techniques but to keep abreast of their runaway multiplicity.

The telephone alone—at first hardly more than a middle-class conceit—now leads the field. Direct dialling was until only a short time ago confined to restricted local areas; by the early 1970s most of Europe and the United States were linked by direct dialling. Complete world coverage is not far off.

Not only has the telephone opened person-to-person contact, it has moved in as an information medium. In most countries the subscriber may dial for the time; in many countries the service is wider: tourist information, sports results, recipes, motoring advice, share prices, weather reports, gospel readings, pop recordings—all these are available by dialling. In Switzerland the caller may dial for news of avalanches.

The telephone system has developed into a communications network infinitely more versatile than could have been envisaged by the pioneers. Apart from speech, telephone lines today serve to transmit other services as well. Users of computers may now feed data to the system, setting up complex inter-city information networks, operating data banks, and 'plugging-in' to specialized computer facilities anywhere in the world.

Documents, instead of passing through the conventional postal system may be transmitted within minutes over the telephone. Dialling the distant number, the caller clips the document to a drum. He sets the drum rotating. A scanning head records the light and shade of the document as it turns, translating intensity of tone

The telephone is vital to the motorway concept. Originally provided only as a special facility for motor club members, telephone points like this now offer all motorists emergency contact throughout the network.

When Grahame Bell visited Europe in 1878 his invention roused little interest. By the early 1970s, though production and installation figures continued to rise, demand greatly exceeded supply. In Britain there were twenty-three telephones to every hundred people. In America, where the proportion was fifty-four to one hundred, telephone conversations reached a national average of some 240,000,000 a day.

into electrical impulses for transmission over the wire. A comparable instrument converts the impulses back into a copy of the original at the other end. Total transmission time: four minutes.

Pictures too can be sent by similar equipment. Press photographs, police portraits, plans, drawings and diagrams—all can be channelled through the public telephone system.

For live radio interviews and on-the-spot eye-witness accounts the nearest telephone provides immediate contact. For telephone conferences, with perhaps a dozen men in as many countries, telephone users may arrange a multiple hook-up—an international consultation without leaving their individual offices.

Telephone lines carry the telex services too. Again, by dialling a number, the user's typewriter is placed in instant contact with one or more elsewhere. Typewritten interchanges, all permanently recorded in black and white, may take place across the world.

In its earlier years the telephone was tied to a system of wires,

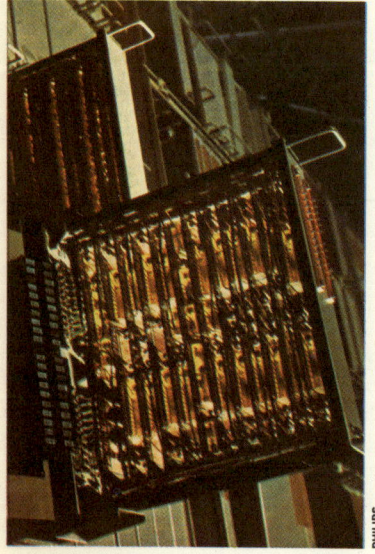

The electronic telephone exchange, miniaturized, and free of the mechanical hazards of conventional equipment, offers new freedom to the communications engineer. Electronic components, housed in push-in racks like this one, provide instant response, one-button dialling for a selected range of frequently used numbers, and self-actuated fault locations.

The cordless telephone switchboard (above, right) is the electronic up-dating of the earlier multiple cable board. As well as obviating plug-in wear and tear, the new-style switchboard allows automatic 'queuing' of incoming traffic; instead of receiving the 'busy' signal, calls are received and accepted in rotation.

first above the streets on poles, then underground in conduits. Today the telephone has begun to break free. Microwave transmission techniques allow countrywide transmission of telephone calls through the air. Transmitting from point to point, tall towers now beam as many as one thousand five hundred telephone calls on a single carrier wave. In addition, television transmissions may go by the same means; the world's microwave towers, linked with each other and with communications satellites, complete a circuit of contact almost too comprehensive to grasp.

However much postal codes and other devices speed up the mails, it seems clear that letters, in their traditional sense, are on the way out. The telephone, in every sense, is on the way in. Whatever else it may do, the post office of tomorrow will carry little in the way of mail.

BRITISH POST OFFICE/RICKARDS

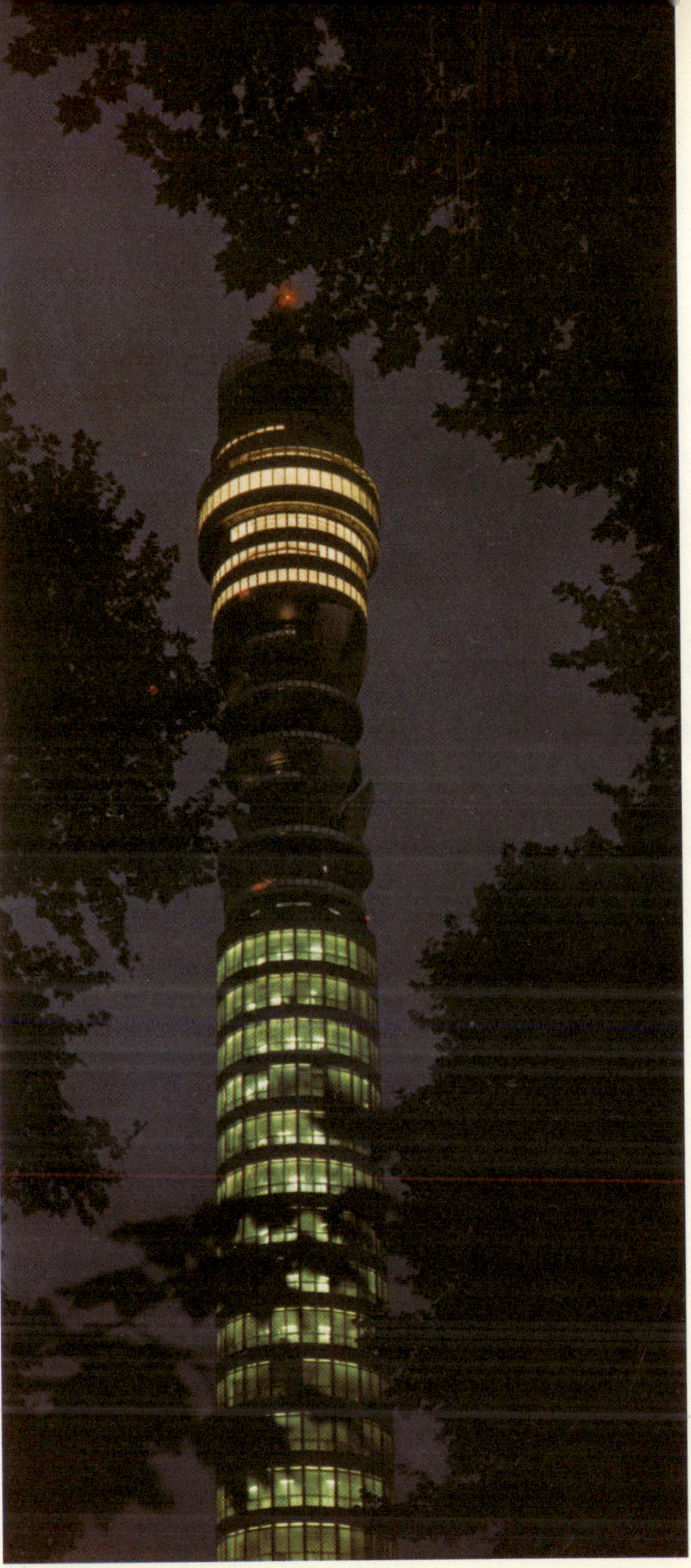

To cut back on the proliferation of cables, telecommunications engineers now make increasing use of microwave transmissions. These, although they have the disadvantage of needing a clear 'line of sight' from point to point, can carry vastly increased information loads. As many as eighteen hundred telephone conversations can be borne on a single carrier wave. Television programmes too can be shuttled from

BRITISH POST OFFICE/RICKARDS

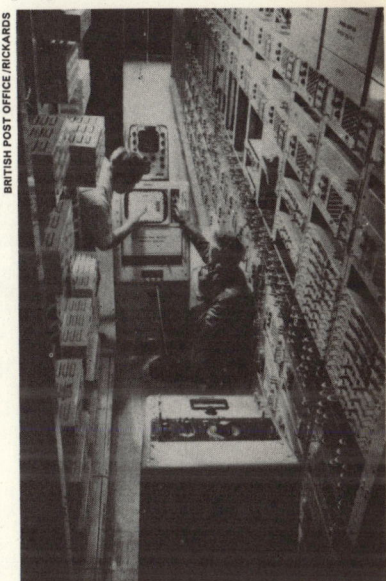

point to point and networked over specific areas. The line-of-sight limitation has produced transmitting towers of great height; this one in London (left) has a potential load capacity of one hundred and fifty thousand telephone calls. Alternatively it can carry one hundred two-way television channels. Much of the associated equipment is contained in the body of the tower. In the smaller picture, taken at the base of the tower, television engineers check the response of the vision circuit amplifier on test equipment in the TV repeater station.

NEWS CIRCLES THE GLOBE

Speeding the Flow of Words and Pictures

IT TOOK sixteen days for news of the Battle of Trafalgar to reach London; America heard of Napoleon's defeat at Waterloo just ten days after it had happened. Even in more recent times newspaper readers accepted a delay of days in the arrival of stories from long distances. In 1916, when the *Titanic* struck an iceberg off the coast of Newfoundland, it was two days before the full story of the disaster appeared in the New York newspapers. London got the full story three or four days after.

Since newspapers first got into their stride, newsmen have struggled to shorten the gap that separates the happening from the report on the printed page. Some of them have devoted their whole life to this single task. One of them, founder of what has become a world-famous institution, was Julius Reuter.

He started with pigeons, flying stock market prices from Brussels to Aachen. He beat the train service by seven hours. Then he got interested in the electric telegraph, at that time still a doubtful starter. Hearing of the laying of a telegraph cable between Dover and Calais, he decided on London as the headquarters for his stock market news service. Settling in London, he added general news to market quotations; he offered news to those who wanted it most —the newspapers. Above all, he offered *accurate* news, *quickly*.

Speed became a watchword. When the American Civil War started, Europe's appetite for news rose sharply, but the mailboat service was slow and rigidly scheduled. News despatches were landed at Queens Town, Ireland, and taken twenty miles back on their tracks to Cork for telegraphing to London. Reuter stole a march on the normal mail delivery. Setting up his own telegraph wire from Cork to Crookshaven, on the westernmost tip of southern Ireland, he arranged for his own messages to be pitched overboard and picked up by special launch as the ship rounded the point, many hours before it reached Queens Town. From Crookshaven they were telegraphed to Cork, and then onward to London. They beat their competitors, often by as much as twenty-four hours.

A Reuter correspondent, actually present in the theatre when Abraham Lincoln was shot, wired the story to New York, where another Reuter man rushed it to the mailboat departure quay. But the boat had just left. The next sailing was in a week's time.

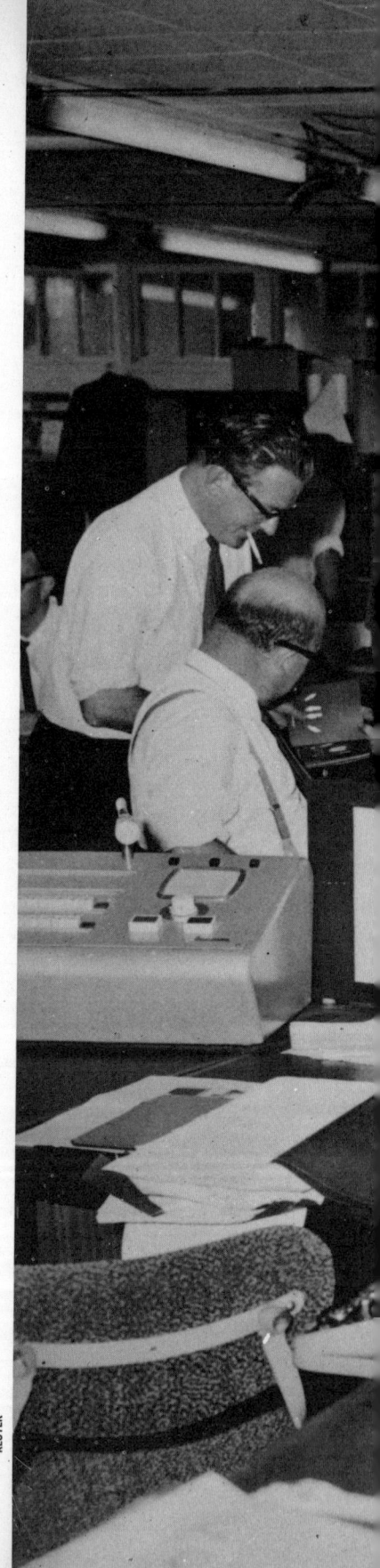

Distribution of news is a round-the-clock service. The world's agencies, clearing-houses for the raw material of press, radio and television, are hard-boiled, busy, informal and efficient. This is part of Reuter's office.

Competitors resigned themselves to waiting a week. Reuter's man hired a fast boat, polished off the story as he chased the mailboat, and eventually caught it up. It would appear, from his shouted instruction as he tossed the message in its container to an officer on board the mailboat, that the unofficial mail drop was an understood routine: 'Here you are,' he said, 'Crookshaven as usual!'

If there was something of the swashbuckler in the Reuter approach, it was not without cause. Competition in the handling of news had become fierce. The world's newsmen wanted to give their readers not history but actuality.

Reuter was by no means on his own as a wholesaler of news. Paris had its Charles Havas, Berlin its Hermann Wolff. In their respective areas their names were soon to become household words. With a multitude of journals as their customers, these men were able to spread the cost of expensive news-gathering. Each individual journal was offered a news coverage far in excess of anything it could have afforded on its own. The smallest provincial newspaper, by subscribing to telegraph services, became part of an international network.

Speed was a major factor. So was accuracy. There was also the question of impartiality. From the beginning Reuter realized that, in the long run, the race would go to the agency that the world's press could trust. While other organizations fell under the sway of political or national influence, Reuter maintained a rigid independence. As his organization grew, he and his associates became more and more deeply imbued with the principle of impartiality. It was to become a guiding light; soon the word *Reuter* at the end of a despatch acquired a special connotation: it was a virtual guarantee of objectiveness and reliability.

Reuters expanded to cover the whole of Europe, then the United States, then India and the Far East, and finally Australia. Through-

Only in relatively recent times have news transmission speeds approached today's high standards. When the steamship 'Titanic' struck an iceberg in the Atlantic on 14 April 1912, the loss of 1,500 lives was a news item that only slowly emerged. Communications failures, like the one that produced this morning-after headline, were sometimes spectacular.

out this period Reuter continued to insist on objectivity. To other agencies he set an example; when, in Britain and the United States there emerged cooperative agencies run by the newspapers themselves, the principle had become universally accepted.

Thus Julius Reuter established not only a world-wide organization, but also an idea. It was an idea fundamental to the free flow of news. In the 1940s, to ensure that Reuters would never become the instrument of any party or faction, it was incorporated as a Trust. Now, under joint press ownership, it serves as a non-profit-making international news channel, completely free of the pressures of nationality, politics or financial interest. Julius Reuter's idea has taken root.

In the 1970s the Reuter network is one of the world's most advanced communications systems. Through the electronic key-

There are over three hundred Reuter's correspondents at key points all over the world, each feeding news into the international network. Backed by some eight hundred part-time reporters, these men represent a fulfilment of Julius Reuter's vision of a truly independent news service—objective, reliable and fast. There are Reuter's offices in most countries; pictured here are (far left) the New York office (centre) Bangkok and (above) London.

By means of computerized message-handling equipment, Reuter's head office sorts, sifts and channels the flow of news. Incoming messages, coded for destination and priority, are retransmitted, recorded and stored as required. Selective switching allows a news story to be filed in New York with instantaneous reception in London, Singapore, Tokyo and Melbourne.

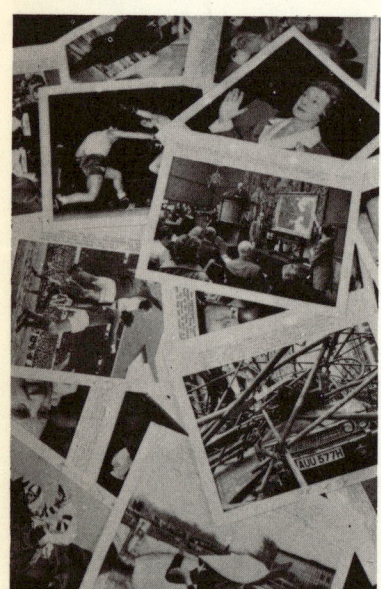

News agency picture transmission, in the late fifties still a novelty, is today a routine operation. The picture (right, above) shows a Japanese transmission office. In the picture below a Press Association picture technician receives a picture on a machine in London. Not only the exceptional picture but the ordinary news shots of film stars, footballers and politicians are regularly shuttled round the world. The spread of photographs in the picture above is typical. Transmissions earlier required a close telephone contact between operators; today they may be sent and received virtually without supervision.

boards of its many thousands of remote-controlled typewriters it provides a virtually instantaneous world news service. Automatically routed to desired destinations, news is flashed instantly from one side of the globe to the other. Simultaneously fed into scores of different cable links, news reaches London, Singapore, Tokyo or Melbourne at the very moment of transmission. Details of an Apollo blast-off, a presidential election result, an assassination or an earthquake, are world property in a matter of seconds.

High-speed transmission is backed by techniques of high-density cable loading. Ordinary telephone cables carry many simultaneous strings of information. A telephone circuit between London and Montreal carries twenty-two separate typewriter channels. Another transatlantic circuit carries the equivalent of forty-four channels.

To handle the distribution and traffic control of messages received for forward transmission at terminal points, computers have been brought in. Without them the avalanche of words that passes through Reuter's main offices would now be unmanageable. Automatically coding, storing and channelling individual items in the flow, the computers perform the 'sorting' process familiar in postal transmission—here, however, in millionths of a second, and without the movement of a single object from A to B.

The Reuter system now incorporates a network equivalent to some four million kilometres of teleprinter link. In addition, there are nearly fifty radio teleprinter links, as well as a New York/Buenos Aires news satellite circuit.

The work of Reuters and other international agencies is complemented by the major national agencies. These, again largely owned and controlled by newspapers themselves, feed national news into the international network and distribute international news within their own territories.

The scene in a big news agency is essentially the same everywhere in the world. At one section news is received; at another it is 'processed' for convenient handling; at another it is sent on its multiple outward journeys.

In the receiving section sit numbers of 'copy-takers'. These 'news typists', wearing telephone headsets, type news straight on to a machine as it is dictated over the telephone by reporters in the field. Each receiving position is served by 'call-dispensers'—miniature switchboards that signal each incoming call; any individual call may be accepted at any one of the receiving positions by any typist free to take it.

The message is typed on a specially prepared form; this gives one master and three copies for each message. The completed story is carried from the typing area on fast conveyors—one copy to a central editorial point for expert assessment of news-value, another to a newsroom for possible development as a major story, another to a section dealing in news for specialized outlets, and another for filing on record. In a few seconds the story is thus in the hands of a number of different specialists, each able to channel the item to best

The Press Association, Britain's newspaper-owned and controlled news agency, takes news from Reuter's and other international agencies and in return feeds British news to the international network. In the picture, copytakers type correspondents' stories as they are phoned in from all over the country. A conveyor routes stories to strategic points for onward handling.

This console, nerve centre of the Press Association's picture transmission service, controls picture traffic through the organization's Fleet Street office. Similar in concept to the route and signalling display in a railway control centre, the unit shows picture channels in use at any moment and allows switching and linking to selected receiving points in Britain or abroad.

advantage; sports news, which would be a wasted item for financial editors, goes only to sports desks; news of merely local interest goes only to that area—and so on. Here again there is the analogy with the sorting-office technique of the mails.

Speedily processed by editors and journalists for its outward journey, the story is finally passed to the appropriate transmission section, either for full scale distribution over all available teleprinter channels, or selectively on given circuits. At a different set of keyboards—this time typing direct from the freshly edited text— transmission typists tap out the news. Unlike the conventional typist, who sees the written words appear on paper as they are typed, these operators work 'blind'; all they see is the message as it is to be typed; for the accuracy of the transmitted message they rely on their own professional care and expertise. Mistakes are rare.

The typewriter keyboard is a major feature of news transmission. Before it finally reaches the keyboard of the typesetting machine in the newspaper printing works a story may pass through half a dozen or more typewritten renderings. As part of the speeding-up process, news technicians seek to keep this number as small as possible; one development is that of feeding news material direct

From the overhead carrier, news stories reach the heart of the Press Association news-handling area. The chief sub-editor (in the foreground), may contact any of the telephone copy-taking points (see picture, page 49) to speak direct to reporters phoning in their stories. It is from here that edited stories go to be tapped out on the keyboards of teleprinters linked to the national network.

from news agency keyboards to composing rooms. By this method it is the typist in the central agency who sets up the columns of the newspaper—not the composing room staff. For tabulated material such as stock market prices the system is already widely used. For general news it is on the threshold; soon it will be widespread.

In their picture sections the agencies provide a parallel service, channelling news photographs by landline and radio to newspapers throughout the world. Here, with the addition of a caption (transmitted with the picture as one image), photographs are shuttled and shunted in a matter of minutes, appearing simultaneously in newspaper picture departments in Tokyo, New York, London, Paris, Edinburgh—wherever the button is pressed on the route-control consoles.

Only a few years ago newspaper pictures transmitted in this way were published with the note 'picture by wire'. Today the system is a matter of routine; no such note appears. To Julius Reuter and his colleagues, with their pigeons, their unofficial mail drops and telegraph wires, such professional unconcern would appear to verge on conceit. But their records show that, like their successors, they would have soon got used to it.

The compact console in the picture below controls the 'routing' of stories as they are teleprinted out of the building. Like the picture route control unit on the facing page, the instrument cuts out log jams, guiding transmissions only to destinations requiring them.

SEEING AT A DISTANCE

New Eyes for Science and Industry

ON THE TABLE of the operating theatre lay a twenty-year-old girl, her heart stilled as surgeons worked on it. The operation was for 'a hole in the heart'; the task was to close the gap in the central wall of the heart—a defect from which the girl had suffered from birth.

The quiet of the operating theatre was broken by the low tones of the surgeon and his assistants, by the occasional clink of instruments set aside on trays and trolleys, and by the sound of the heart-lung machine as it took over the function of the patient's own heart and lungs.

It was a routine operation. But it was not quite an ordinary one. In the lower surface of the operating theatre light fitting, hardly noticeable, was a glass disc—the eye of a television camera. In an auditorium, remote from the scene, yet intimately observing it, two hundred people sat and watched.

The year was 1967. The place, the Surgical University Clinic of Munich. It was a unique occasion, not because television was transmitting details of a surgical operation to a distant point; that was now a well established technique. It was unique because two hundred people were viewing it at once on one ten-foot by ten-foot screen—in colour.

While Dr Klinner and his assistant worked in the theatre the viewers saw every detail; they saw the needle pushed first through the plastic 'patch', then through the wall of the heart. They saw the stitches—on the large screen, appearing as thick cables; they saw the gigantic knotting of the threads. They saw, for the most part, somewhat more than Dr Klinner himself.

When the plastic patch had been finally stitched into place and the heart closed again, there came the moment for the heart-lung machine to be switched off, and for the patient's own heart and lungs to come back into action. The auditorium was hushed as electrodes were applied to the heart to stimulate it to beat again. Uncertainly it fluttered into life; soon there was restored the rhythm of the normal beat. Two hundred people, holding their breath as they watched, gave a sigh of relief.

For the audience, this had been a television programme more impressive than any on their TV screens at home. For scores of subsequent audiences, at this and other such centres, television

In its ability to provide instant close-up vision, television has become the world's most powerful teacher. Its role in education grows daily. In the picture, poised for action, a classroom chemist awaits further instructions.

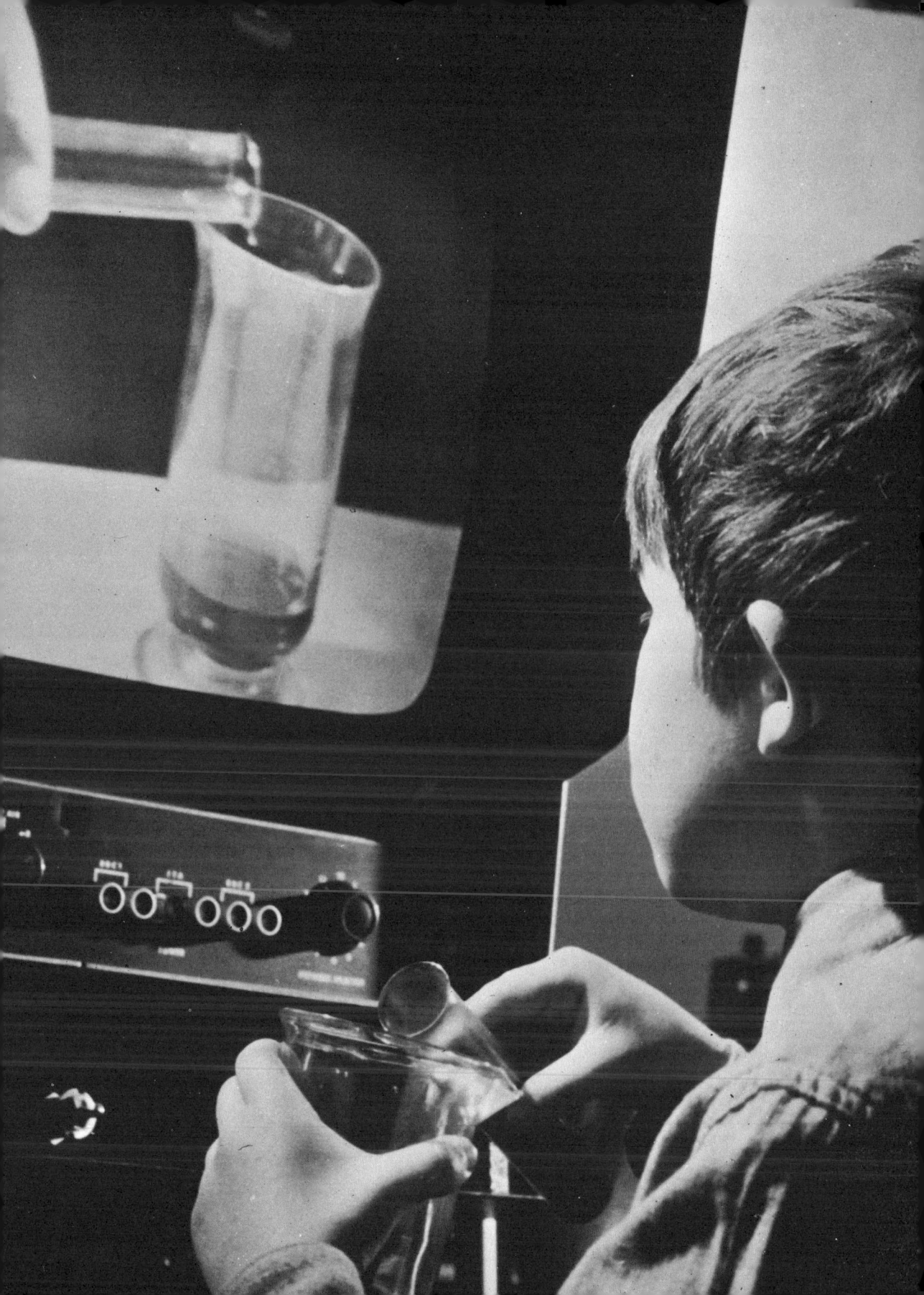

has taken on a role more significant than many had thought even remotely possible.

Television is rather less than a lifetime old. Its development, from a vague shadow on a screen in a Soho attic in 1924 to the universal image of today, has transformed the world.

Though first publicly demonstrated by John Baird in 1926, the idea had been brewing for some time. In 1873 a cable operator in Ireland discovered—by accident—that the electrical characteristics of a substance called selenium were affected by the action of light. From that moment physicists all over the world realised that here was the beginning of the transmission of images. An effect that could be induced as a crude reaction could be refined... What was more, if electricity could travel over distances, as telegraph messages did, then the refined effect—light and shade—could be sent from point A to point B... Television!

How to achieve this refinement of the effect? It was a good

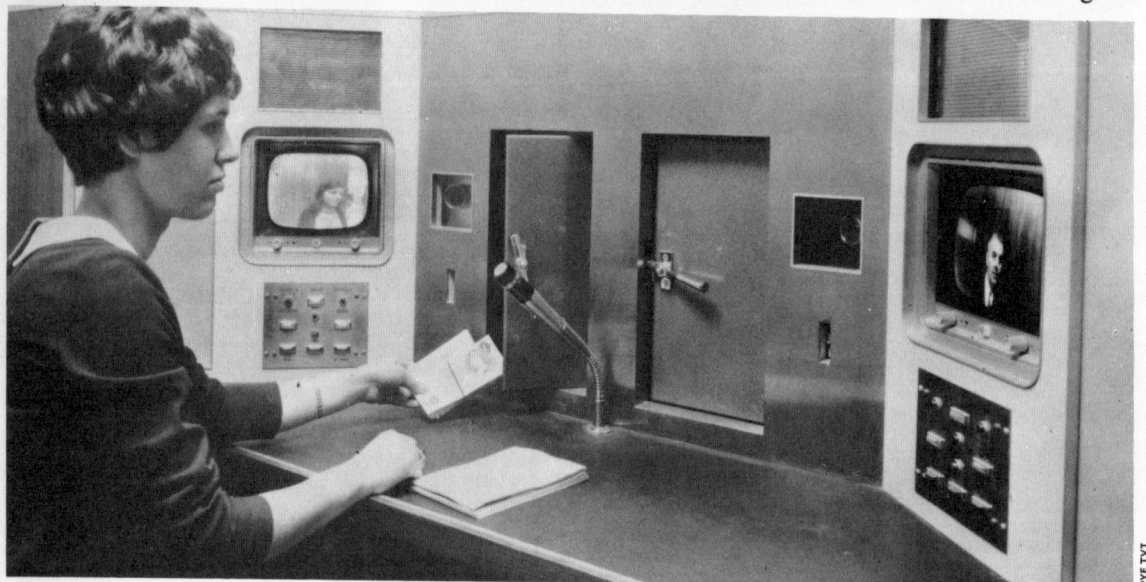

At a bank in south-east England a counter clerk on the first floor serves customers in cubicles on the ground floor. Air tubes carry cheques, statements and notes between them; microphones and two-way closed-circuit TV provide the personal link.

question. It was like conceiving the possibility of a blast furnace from the presence of a single spark.

By the turn of the century pioneers in a dozen countries were hard at work. Most of them were unaware of the activity of the others. Each pursued the problem according to his experience, his facilities and his personal hunch. Whether they knew it or not, all concurred in one respect; the image to be transmitted must first be translated into simple electrical impulses and afterwards reconstituted to form a duplicate image at the receiving end. The key question lay in the method used in the process of translation—and subsequent retranslation.

There began to emerge two broad approaches; one was based on the magnetic manipulation of a stream of electrons in a vacuum tube, the other relied on the mechanical 'shutter' effect of a rapidly

rotating perforated disc. Both of them appeared possible starters.

It is one of the ironies of history that the approach that was in the long run destined to fail was first in the field. John Logie Baird, ailing and headstrong son of a Scottish minister, set himself the task of achieving a true transmission of vision. He tried it the hard way. He placed his faith, his health and all of his slender means on the outside chance—the mechanical disc system.

In his Soho attic, with apparatus that was literally tied together with pieces of string, he worked on the problem. Short of money, often short of food, even short of sympathetic interest from friends and neighbours, he plodded on.

On 27 January 1926 he invited fifty experts to watch a demonstration. From one room to another in his makeshift place in Frith Street he transmitted a simple picture. It was a crude thirty-line image—a travesty of today's 625-line picture—but it worked, and it was his—John Logie Baird's.

Baird went from strength to strength. He developed first colour and then stereo television. He made the first TV recordings, impressing transmitted signals on waxed discs. He transmitted television from Britain to a ship in mid-Atlantic and later from Longacre in London to Hartsdale, New York—a distance of four thousand miles.

But it was to be the other approach that won the day. While Baird forged ahead with his disc system the cathode ray tube was being developed; when Britain's broadcasting authorities came to consider a regular public service of television the question had to be decided: which was it to be, the Baird system or the newcomer's? After a period of testing they made their choice. In early February 1936 the world's first regular service of television went over to the Marconi-Emi electronic system.

Britain's TV audience at that time was about ten thousand. Today, an average viewing figure is ten or twelve million. A

Television contact now offers conference facilities to delegates sitting in their own home towns. The 'confervision' link, a visual version of the long established 'telephone conference' facility, is a rapidly expanding service of Britain's Post Office. Ultimately it will be possible for delegates to participate from individual offices.

This tubular TV camera, impervious to radioactivity, is designed for remote inspection of atomic reactor piles. It is one of a wide range of special-purpose cameras used in science and industry.

55

Eurovision programme, shared by a dozen countries, may reach some hundreds of millions. From being a minority fad in the middle 1940s, watching television has become one of the most widespread and time consuming of all human occupations.

In July 1969, just two years after the big screen surgical transmission in the Munich hospital, there was another big screen showing. This time it involved big screens all over the world, and this time transmission distance was greater. An estimated audience of six hundred million people watched images from two hundred and fifty thousand miles away.

The television camera serves as a universal watchman. This one observes the state of the traffic in a Tokyo street. Remote-controlled camera mountings allow movement of the camera from side to side and up and down, zoom-lens facilities, and colour filtration. Some externally mounted cameras are fitted with remote controlled 'windscreen wipers'.

'I hope you are watching,' said Neil Armstrong on the surface of the moon, 'I hope you are watching how hard I have to hit this into the ground.' He rammed a core-sampling rod into the lunar surface. A major portion of the earth's population craned closer. They were watching all right.

The camera, specially manufactured for the mission, was designed to be held by the astronaut or mounted on a stand. It weighed just over seven pounds and sent back its pictures on a channel which it shared with voice, biomedical and other scientific data transmissions. Inside the spacecraft a second camera transmitted television pictures in colour throughout the flight. The viewfinder of the second camera consisted of a miniature television screen measuring two by two and half inches; it showed the exact scene as received on earth.

Space travel was still a wonder, but to the millions who shared in its excitement through television, the TV camera and the TV screen had become as familiar a feature of everyday life as the telephone, the pocket radio and the tape recorder.

In entertainment and news transmission television had become the dominant medium. By the end of the 1960s it was also a factor in education, industry, research and technology at large. The moon transmissions were a culminating demonstration—half news, half technology—of television's role as an instrument of our time.

As the 1970s opened the television camera had sought out and solved almost all of the 'seeing' problems of organized society. Today there are few fields where the TV eye does not operate. Whether as watchdog, teacher, meter reader, weather-eye or traffic cop, it enormously extends the limits of human vision.

Television cameras, remotely controlled to watch traffic conditions in crowded streets, are installed in a dozen of the world's great cities. From central headquarters traffic experts smooth the flow of people and vehicles. Multiple screens before them show the conditions at key points in the network. The experts add visual oversight to computer control. A flick of a switch brings a new picture to one of the screens; a lever is moved and a distant view zooms smoothly into close-up; with a touch of a button a camera angle moves through ninety degrees; a cross-roads view now shows another aspect of the traffic flow. Buttons are touched, interconnected traffic lights change from red to green, green to red.

In a major industrial plant a night-duty operator sits at a desk examining dials and meters in remote parts of the works; a twelve-way switch allows him to flick from one to the other at will, entering readings from each in a progress log; he covers acres of ground without moving from the spot.

In an institute of aviation medicine, astronauts are strapped into their seats in a training centrifuge—a device designed to simulate extremes of gravitational force; focused on each is a TV camera; from central control biologists observe the subjects' reactions as

Paris is one of the many cities now equipped with television traffic control; the picture was taken at the formal opening of the traffic centre in 1968. Control consoles in the foreground are used to adjust camera angles and to bring alternative cameras in different parts of the city into vision on the monitor screens. Computerized traffic flow systems, in experimental use in the sixties and early seventies, use television oversight for operational assessment.

In the 'hot laboratory' of an atomic reactor (below), TV cameras give close-up views of remote-handling operations. 'Slave-manipulators' reproduce the operator's finger movements in the danger area; TV screens show larger-than-life-size detail.

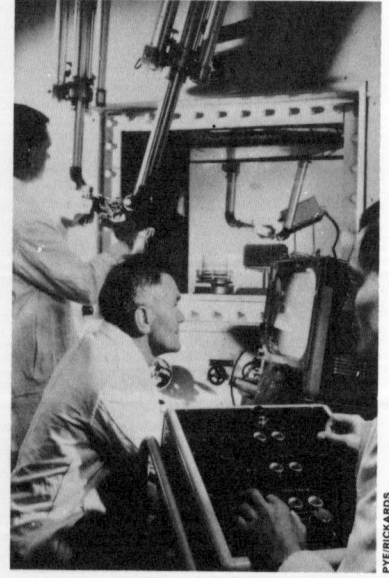

the giant centrifuge spins its passengers ever faster.

In the operations control room of a salvage ship a TV screen glimmers, first softly, then bright and clear; the men watching see the steel plates of the hull of a ship slide by; floodlights on an underwater TV camera pick out every rivet. Under telephone instruction from the control room the diver floats with the camera along the flank of the sunken wreck; soon an aperture in the plates appears; it is here that entry will be gained; soon the actual salvage operation will begin.

In the radioactive area of an atomic reactor installation, mechanical arms manipulate dangerous substances; there is no human present but the arms behave with uncanny confidence; they stretch, pick up, put down, with the skill and dexterity of human agency. A mechanical thumb and forefinger picks up a beaker containing a liquid, takes a test-tube from a rack and carefully pours the liquid into the test tube—just so much and no more, without spilling a single drop. Focused on the scene is the

zoom lens of a remote-controlled TV camera; fifteen feet away, behind many tons of protective concrete a scientist operates a pair of master arms, each movement of which is duplicated in the radioactive danger area. It is this man who 'sees' the distant beaker and test tube on a stereo television viewer; his contact with the dangerous operation is as complete as it would be if he were in there on the radioactive side of the concrete.

Seeing at a distance is just one aspect of man's increasing ability to act at a distance. It is becoming easier to go from A to B without, so to speak, moving from the spot. The fifteen feet that separates the atom scientist from the subject of his actions could as easily be twenty feet, or two hundred feet, or two hundred miles.

It could as easily be the distance from earth to the moon. With distant vision to guide him and remote manipulation to effect, it looks as though man's influence on his surroundings is due for a rapid expansion. It may prove to be a step as significant as his first acquisition of speech.

Large-screen television and computers linked by satellite were features of a transatlantic computer experiment set up at Britain's National Physical Laboratory. Results of the experiment, registered on computer read-out at the British terminal, were shown in close-up by closed-circuit cameras. Giant enlargement of the TV image appeared on the Eidophor screen. This picture, taken during an intermission, shows a speaker (left of centre) at the rostrum, his TV image (left) monitored in the back of the camera focused on him, and (right) his magnified image on the Eidophor screen. The screen can enlarge a TV image up to one hundred by sixty feet.

NOW, FACSIMILE AND VIDEO

Vision by Wire, Tape and Disc

IN THE PICTURE TRANSMISSION ROOM of Tokyo's leading newspaper, *Asahi Shimbun,* a technician pressed a button. The scanning cylinder of the transmitter began to roll. Five hundred miles away, in the city of Sapporo on Japan's Northern Island, another cylinder turned simultaneously; this second cylinder, the receiving drum of the transmission link, carried a wrapper of photosensitive film; as the scanning head of the Tokyo cylinder moved slowly across the rotating image the film at the Sapporo end recorded the response of the scanner. As on a hundred previous occasions, an image was being sent through the air.

Picture transmission—both into and out of Tokyo—is a well-established drill. Ordinarily there would have been nothing to call for comment. But on this occasion telephone lines between two offices were busy on one subject alone—the progress of the transmission. This time, instead of sending just a single picture, the *Asahi Shimbun* was sending the whole of its front page. And not only that: the newspaper was transmitting each following page in quick succession. For the first time in history an entire newspaper was being sent by radio for plate-making and virtually simultaneous publication in cities five hundred miles apart.

Within a few minutes the film of the first page was off the Sapporo machine and being rushed into the plate-making department. By the time page two was ready to come off, the first completed plate had been cooled and clamped into position on the rotary cylinders of the printing press. Everything went without a hitch; with the last of the pages successfully transmitted and the last of the printing plates completed, the presses were ready to run.

Only a short time later thousands of copies of the paper were on sale in Sapporo. The people of Japan's Northern Island, cut off from mainland Japan by sea and rail journeys amounting to anything up to forty-eight hours, now had a same-day Tokyo newspaper. The button was first pressed on 1 June 1959. Today, with improved production techniques and streamlined distribution the *Asahi Shimbun* is on sale in Sapporo only one hundred minutes after the same edition appears in Tokyo.

Whole-page transmission of newspapers is becoming not just a parlour trick but a vital information link. Australia's national

The moving drum of the facsimile machine is a symbol of the age. Like 'video'—magnetic storage of TV pictures—it typifies the concept of electronics as substitute for physical transportation of messages from A to B.

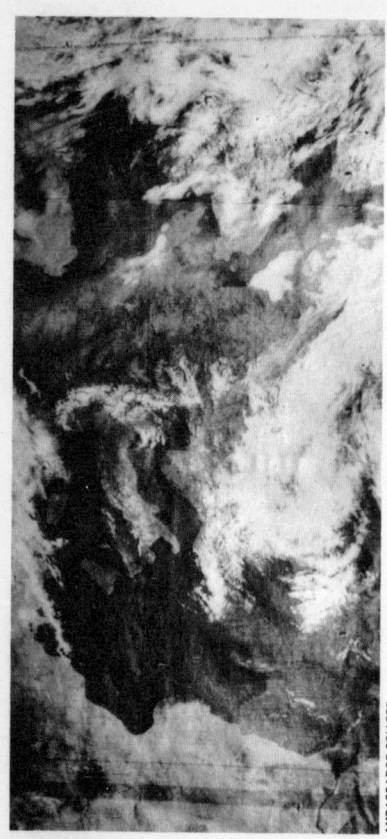

This television image was made from six hundred and ninety miles above the earth's surface. It was shot on an exceptionally cloudless day, and was transmitted to earth by satellite Nimbus III *as part of its normal weather-watch. Britain appears at top left, Italy and North Africa in the lower area of the picture. Transmissions from the satellite are fully automatic, operating without the aid of manned ground stations. Pictures can be received (free of charge) by anyone who cares to build the comparatively simple receiving equipment; signals are converted into pictures by the use of a modified commercial facsimile machine.*

newspaper, *The Australian*, printed in Canberra and Melbourne, was earlier bedevilled by road and air transport problems; printing plates had to be freighted through fog, snow and other hazards to a tight production schedule. Today, regardless of weather and traffic, the whole newspaper goes over the telephone line at an average speed of five minutes per page.

Newspapers in a dozen countries have adopted the method. To show its paces, the London *Daily Express* transmitted the whole of its front page to a conference of news communications experts three thousand miles away. The conference was in Costa Rica; the transmission was by satellite. When colour pictures of Neil Armstrong's moonwalk reached New York the same newspaper transmitted them direct to its London printing office, again by satellite, *in colour*.

In the early 1970s some newspapers were beginning to extend the principle into the reader's home. Two Japanese companies had produced television sets incorporating facsimile newspaper page receivers. During the night and early morning, when the television is out of ordinary use, its aerial and receiving channels operate a domestic page receiver. The *Asahi Shimbun* emerges ready printed from a slot at the side of the TV screen. A second company has developed a radio newspaper independent of the TV. This is a simple domestic version of the big facsimile machine.

Facsimile transmission is the perfect expression of a communications ideal; movement of information from A to B—with no physical handling of three-dimensional objects, no packages, no trucks, no tracks. Like the telex typewriter, facsimile transmission provides a visible record for the recipient to study. But it goes one better: it allows transmission not just of typewritten characters but of visual material of virtually any kind.

Facsimile networks now link offices all over the world. Meteorological offices send weather maps direct to ships at sea; police forces exchange pictures, fingerprints and criminal records; printers send proofs to customers for approval; engineers transmit plans, requisitions, modification details. The flow of pictures increases daily. Most of it is carried over ordinary telephone wires. Some of it goes over the air: police radio cars can receive identikit pictures while actually on the move. All of it, by wire or by radio, is free of the danger of error in transit. The most complicated of documents, chemical formulae, foreign languages—even doctors' prescriptions—are transmitted in faithful detail.

In large-scale international ventures, where groups of specialist companies co-operate in joint planning, the facsimile machine is a vital tool of trade. The case of the Anglo-French Concorde operation is typical. Here, as though within the same building, French and British engineers working in Toulouse and Bristol were able to compare notes as the joint operation progressed. Backed by voice and teleprinter circuit, a two-way facsimile link provided minute-to-minute contact.

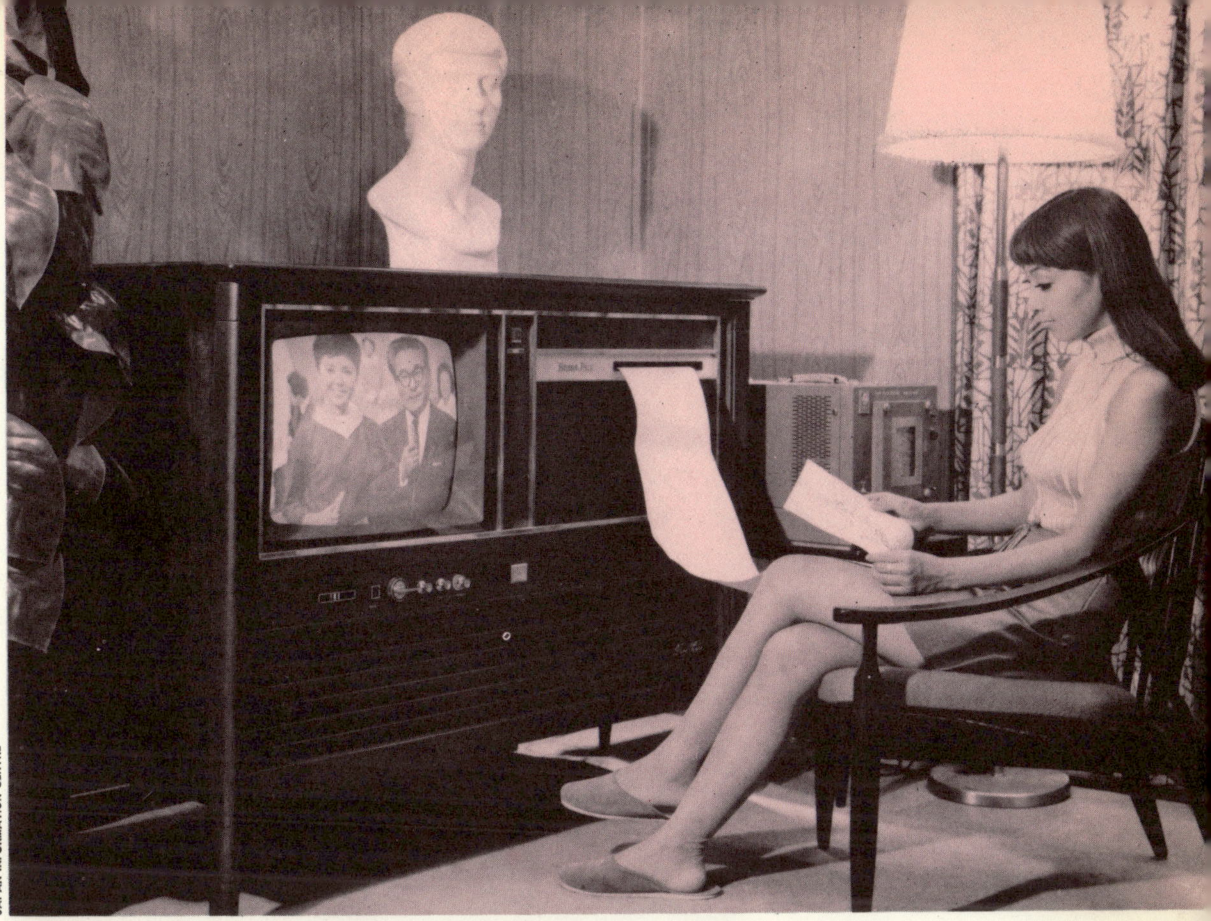

JAPAN INFORMATION CENTRE

MUIRHEAD

The facsimile machine can combine with the domestic television receiver to provide a permanent record of news and other items appearing on the screen. In Japan, where this picture was taken, one daily newspaper experimentally transmits its front page overnight while ordinary TV channels are out of use. The girl in the picture is looking at a special news and weather sheet, part of another pilot service transmitting simultaneously with TV broadcasts.

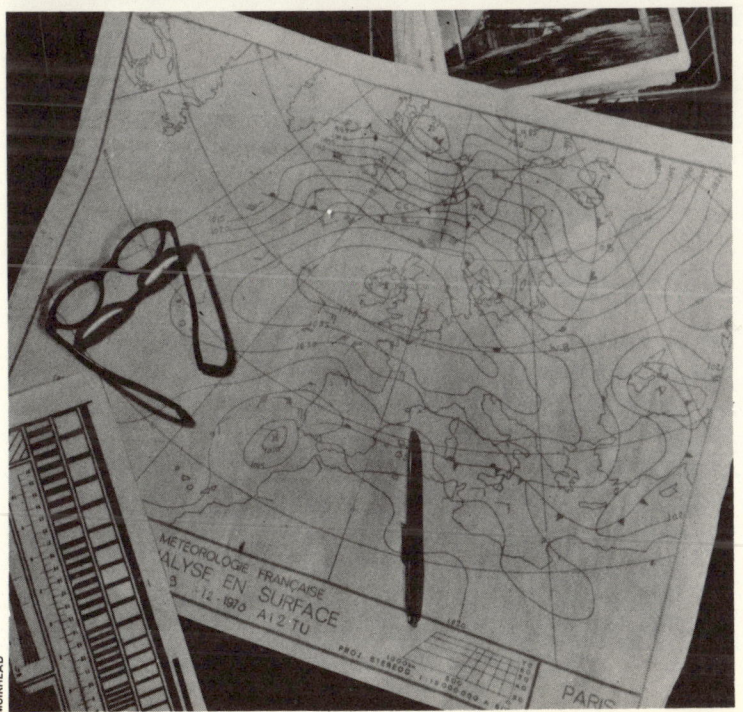

Weather charts, prepared and transmitted by meteorological services, give up-to-the-minute weather situation reports, transmitted by landline or radio and received on facsimile machines anywhere within contact.

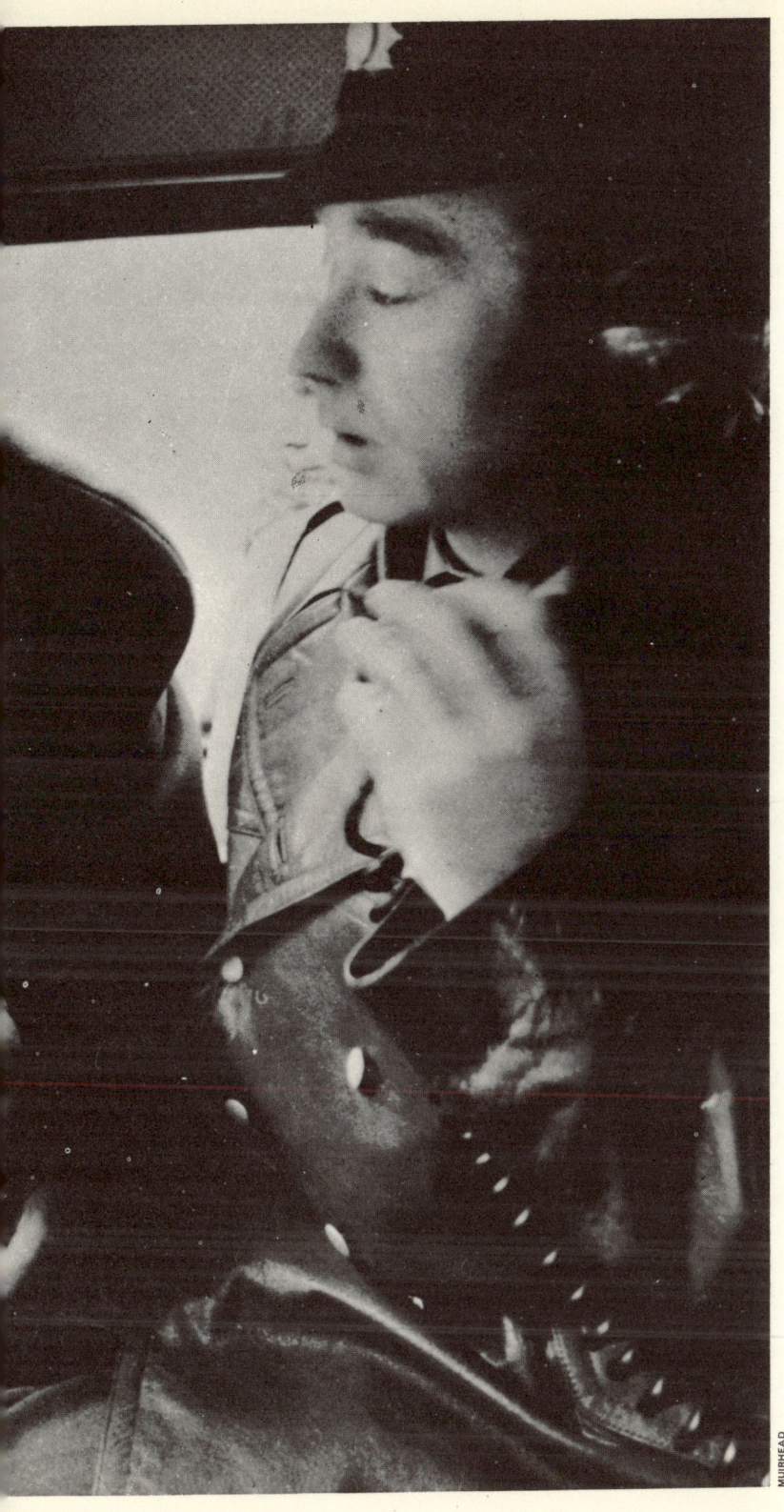

Operational trials by police in Britain and West Germany have shown the value of direct facsimile transmission to patrol cars and other vehicles. Facsimile machines are today a well-established feature of police communications but until the early 1970s their use was confined to static situations. Now, using dashboard mounted receivers, photographs, identikit pictures, and other data are transmitted over the vehicle's normal radio system. In the photograph shown here reception of the picture is being telephone-monitored by the recipient; in later models transmission is controlled entirely by the base station, allowing documents to be sent to a vehicle on the move without distracting the driver, or to an unattended vehicle to await the driver's return. Maps, car registration numbers, details of missing persons can be transmitted within minutes. Document transmission speed, on a paper width of four inches, is four inches per minute. Pictures may be transmitted on a selective basis, sending specific documents to individual cars, or broadcast for general reception by the full fleet. One significant advantage is that the machine is silent—there is no possibility of information being overheard by bystanders—and it will operate in conditions where static or distortion may make speech unintelligible.

Remarkable fidelity of facsimile transmission—even in colour—is shown by the comparison of a colour photograph (left) as transmitted and (right) as received. The picture was sent as a set of three colour-separated negatives, with superimposition at the receiving end. The newspaper reproductions show part of the front page of the Daily Express as transmitted from London by satellite link (above), and as received and printed in Costa Rica (below) on the occasion of a newsmen's telecommunications conference.

In the field of medicine, where complex laboratory reports may need rapid transfer from one centre to another, the system is irreplaceable. In Britain a group of hospitals in the Midlands uses facsimile transmission for blood analysis reports. The results of blood sample tests carried out at the Group Pathological Laboratory in Warwick are transmitted to participating hospitals within a few hours. Some sixty analyses go out each day to twenty different wards. Where previously results were telephoned round in a lengthy session—with loss of technicians' time as well as risk of error—today they reach the wards in exact facsimile by simple insertion in the central transmitter.

Though it is still, in the early 1970s, too expensive for widespread use, the facsimile machine has moved unmistakably into the orbit of the ordinary man: desktop models, hardly bigger than a portable typewriter, are beginning to appear in business offices; before very long they will be as familiar a feature as today's office copier.

Communications experts in many countries see the facsimile as

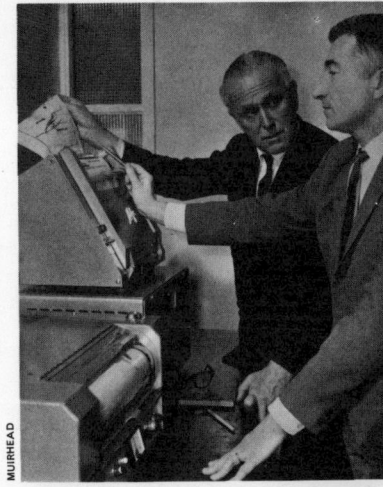

In the picture on the left a hospital nurse stands by to tear off a batch of blood-test analysis results received from a central laboratory in a scattered hospital complex in Britain. Top picture shows the Bristol-Toulouse facsimile link on the Concorde prototype construction operation. The fingerprint picture is part of a full ten-finger standard form transmitted by police facsimile.

the beginning of the end of the conventional postal system. With increasing fidelity of reproduction and ever faster transmission speeds, the facsimile comes into its own. As it does so the postbag becomes irrelevant. Ultimately, it seems, the mails will be as invisible, and as weightless, as telephone calls.

So much for the transmission of the still picture, but how about the moving picture? The television image, until the early 1960s, suffered from impermanence: no sooner had it been transmitted than it had vanished for ever. Techniques were evolved for the making of ordinary cinematograph films direct from the screen, but these were never entirely satisfactory. Of many drawbacks, the biggest was time-lag in processing the film.

What was needed was an entirely new approach—a system as simple and straightforward as an ordinary tape recorder, in which programme material could be taped as it came over the air and, if desired, replayed immediately.

It was in fact precisely this that was developed. In April 1956 an

TV tape recording is a logical extension of the principle of taped sound. This 'pocket TV studio kit' provides a camera, tape-deck and TV receiver. Playback is immediate. The equipment is cartridge-loaded and operates on household mains or battery in black and white or colour.

Tape recording of the television image, itself a recent development, is now augmented by the use of magnetic discs. The disc, shown here under microscopic surface inspection, permits quicker random access than reel-to-reel tape recording.

American company demonstrated immediate playback of moving television images from tape recordings. Exactly two years later the same company introduced a colour version of the device. As always, miniaturization was a big consideration: by 1961 a TV recorder had been produced giving half an hour's playback and occupying no more than one cubic foot of space. Five years later commercial models were available for domestic use. The TV viewer could now tape selected programmes—even, if necessary, in his absence—and replay them at will. He could also take his own pictures with a home TV camera and play those back at will. By the end of the 1960s moving TV pictures were being taped as a matter of ordinary home routine.

Professional TV benefited in a number of ways. Not least was the ability to replay shots of sport and other action pictures only seconds after they had been transmitted—if desired, in slow motion, and even frame by frame, each frame being held as long as required. An obvious virtue was the ability to shoot studio productions at times convenient to those taking part and to transmit them later. This facility offers an additional advantage in the bigger countries, where great distances produce varying time differentials. The television authority is able to put out a 'six o'clock programme' all over the country, even though there may be a difference of some hours between one area and another. The taped programme is released at the appropriate time in each area.

The introduction of the taping principle has revolutionized communications. Tape now carries not only sound and moving pictures but information of all descriptions. Temperature, numerical and alphabetical information, vibration, acceleration—these and many other performance characteristics can be recorded on tape for instant laboratory playback without processing of any kind.

The basic principle is simple; particles of metallic oxide on the tape form magnetic patterns as signals are fed into the system; at playback the patterns are reconverted into the original signals.

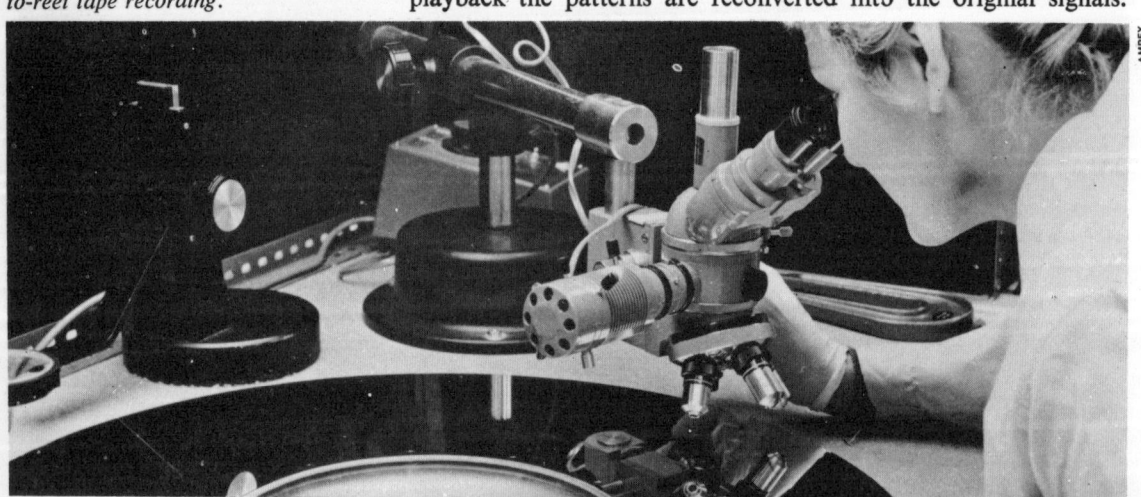

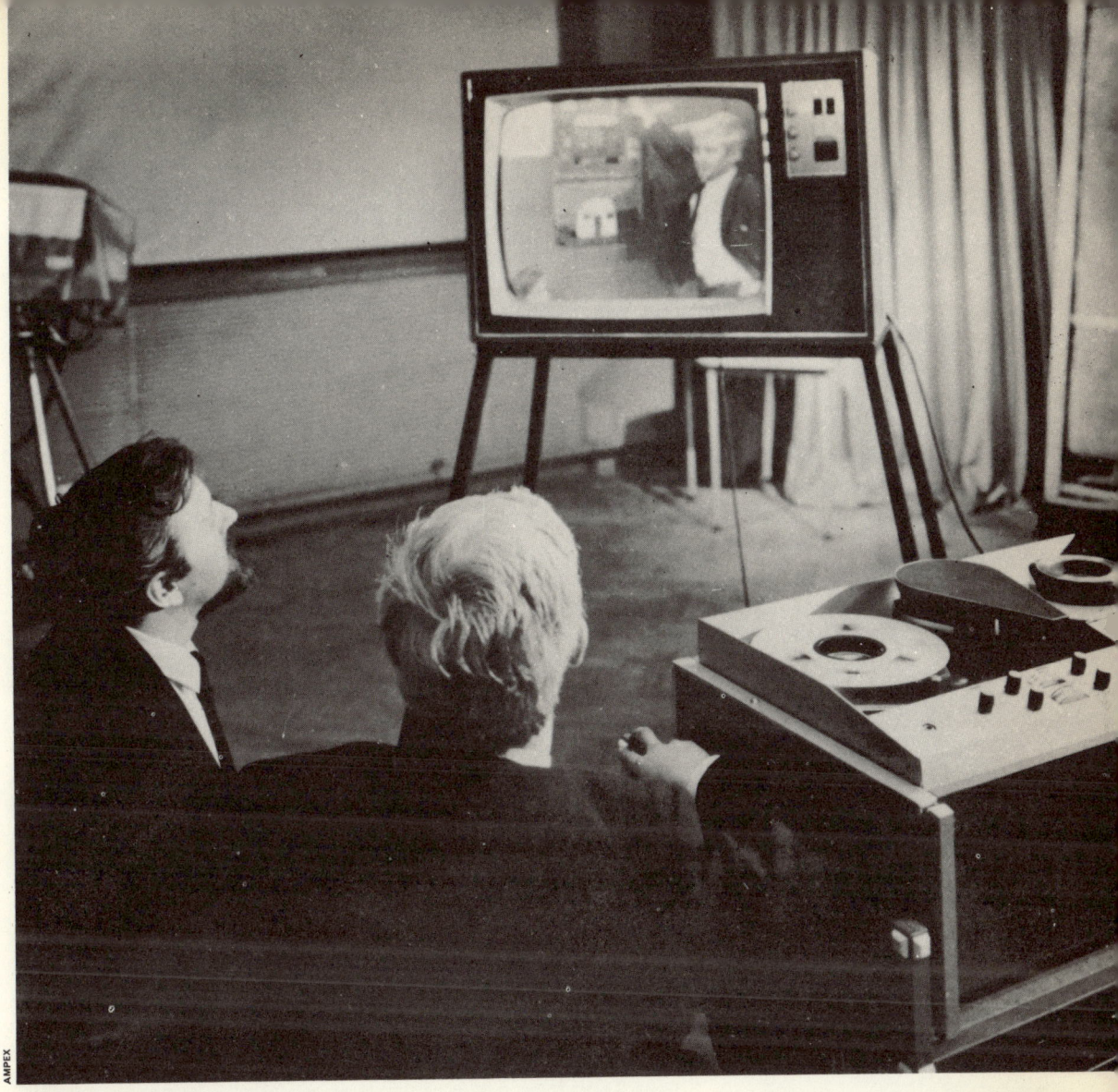

Whatever the nature of the signal coding, the precise sequence can be stored up indefinitely—or sent through the air by microwave to distant points. Statistics, pictures, music, criminal records, typed matter—even whole libraries, can be transmitted. As many as a quarter of a million standard pages from books may be stored on a single fourteen-inch reel of tape; file copies may be viewed on television receivers or as printed reproductions. The operator can file, view, replace, delete or relocate any page electronically in seconds at the touch of a button. Used in association with computer programming, the principle has applications and possibilities almost beyond imagining. One thing is clear: whatever the material, once it is taped it is ready for instant transmission from A to B; the slowest element in the operation is the human mind—first in its ability to take in the information on arrival and secondly in its ability to grasp the implication of its own slowness.

At St Mary's College, Twickenham, a student teacher and his tutor play back a live classroom session, recorded on video tape only minutes earlier, for critical comment and assessment. By seeing himself as others see him the student can more readily correct any faults in his presentation.

MESSAGES THROUGH SPACE

The Role of the Communications Satellite

THERE WAS WORLDWIDE excitement when, in 1966, the Russians brought off the biggest space trick to date: a soft landing on the moon. It was a landing by a machine—unmanned, and one-way only, but it represented a major step in space technology. Instead of crashing to destruction on the moon's surface, *Luna 9* had been gently set down in the desired spot—intact and still transmitting.

Before this date there had been a mounting flurry of moon activity; there had been the first successful placing of an object on the moon (the Russians again, in September 1959, with the moon-shot *Lunik 2*); there had been innumerable spacewalks, fly-by's and unmanned—and unscheduled—crash landings. Now, on 3 February 1966, a man-made instrument sat on its landing pods in the lunar dust and sent back signals to earth.

On earth there were only a few places with radio ears powerful enough to listen. One of them was at Jodrell Bank, where the two-hundred-and-fifty-foot dish of the world's largest radiotelescope had been tracking the spaceship on its journey. Now the great dish was no longer moving. It was fixed on the vital spot two hundred and forty thousand miles away, listening intently . . .

In the control rooms, in the hallways and corridors of the radio telescope building, the atmosphere was tense. For technicians and newsmen alike this was a moment of exceptional significance. As upstairs in the radio laboratory Sir Bernard Lovell and his colleagues analyzed recordings of the lunar transmission, in the ground floor areas news correspondents heard the signals loud and clear. The sound was the typical high-pitched warbling of space-craft information transmission; each rapid note in the sequence carrying its own message of environmental measurement.

But there was another rhythm in the signals, a characteristic not unfamiliar to some of the newsmen present. 'That,' said someone at length, 'is a *picture* coming over—a *facsimile transmission!*' Experienced ears were cocked to the sound from the moon; it certainly seemed possible . . . Perhaps, if a facsimile receiver were brought to Jodrell Bank, the signal could be converted on the spot—the first-ever picture from the surface of the moon . . . The journalists phoning their stories to London broke off and asked to speak to their respective picture desks.

Jodrell Bank, 6 February 1966: 'tuning in' to picture signals from Russia's Luna 9, news reporters previewed the first pictures ever to be transmitted from the surface of the moon, nearly a quarter of a million miles away.

Within minutes the editor of Britain's *Daily Express* had offered the necessary equipment. From the paper's Manchester office came one unit; in London a technician set out to drive through the night to deliver a vital 'accessory'. This, a unique radio link, was equipment specially designed for its own use by the newspaper for transmission of ship-to-shore rescue pictures.

By the time dawn broke the various items were assembled. The radiotelescope men stood by; the facsimile men made ready to feed recorded signals into their machine. But the picture data used in the transmission were still unknown; setting up the receiving

Newspaper telecommunications men, moving in hurriedly with facsimile equipment to Jodrell Bank, monitor picture transmissions from the surface of the moon. One vital piece of equipment, specially developed by the 'Daily Express' for sending radio pictures from sea-rescue operations, was hooked into the circuit to get the lunar landscape pictures.

standards was a matter of guesswork. At the first trial the apparatus yielded nothing remotely recognizable. At the second and third, the same. In the pale light of morning, spirits sank.

Then came an alert from the control room: *Luna 9* had started to transmit again. Quickly the live signals were channelled into the facsimile equipment. Tired men gathered round as the results appeared. For the first time in the history of Man—close-up pictures from the actual surface of the moon . . .

It is an oddity of history that in the reception of these pictures two of the key instruments were used outside their normal function. The facsimile machine itself, pressed at a moment's notice into the service of space communication, had been designed for a potential circuit no greater than that of London/Manchester. The radiotelescope, Sir Bernard Lovell's giant brainchild for Manchester University, played a converse part; designed for the probing of deep space—'for the study of the universe'—as its creator expresses it, the telescope's focus was turned for a change on the homely bleeps of a newspaper facsimile transmitter.

The arrival on earth of the *Luna 9* picture was a highlight in the story of space communications. But there has been much else both before and after. In 1959 Russia's *Lunik 3* had photographed the distant side of the moon. in 1965 America's *Ranger* 9 had sent back in-flight pictures of the moon's surface. In the same year Russia produced another series of behind-the-moon pictures. Within only a few weeks of each other, first American, then Russian astronauts got out of their space vehicles and floated free; their pictures were seen by millions over satellite circuits.

Mariner IV, forgotten by the world at large during an eight-month journey, signalled its arrival at Mars. As it flew past the planet it sent back pictures of the surface. Pictures are expected back from further Mariner class vehicles in the 1970s—some of them from unmanned landings on Mars. Afterwards there are plans for picture-missions to more distant points: Jupiter, Saturn, Uranus, Neptune and Pluto. These, eight- to eleven-year flights, are expected to yield pictures certainly as clear as the latest Mars pictures—possibly clearer.

As the world's attention is focused on long-distance vision, speech transmission has been taken more and more for granted. The impact of the TV picture is in a literal sense so spectacular that the novelty of speech in space begins to fade.

But for millions everywhere in the world the voice capacity of space has immediate relevance. As a communications medium, 'surface radio' has always been less than perfect. Relying heavily on the round-the-world reflective layer of the ionosphere (the invisible sheath that prevents signals going straight off into space), radio is subject to vagaries and eccentricities. Often the reflecting sheath fails to contain the signal and there is fading, distortion, interference—even complete loss of contact. Telephone cables too, no matter how much engineers increase their carrying capacity, have their limitations. What is needed is another medium altogether, more reliable and less crowded—a medium like Space.

In past centuries few innovators lived to see their predictions fulfilled, but this century is different. In 1945 Arthur C. Clarke, science fiction writer, theorizer and visionary, wrote an article in the British magazine *Wireless World* entitled *Extra-terrestrial Relays*. In it he proposed that signals should be 'bounced' from one point on earth to the other through the medium of space.

Clarke's idea was basically a simple one; he analyzed the problem like this: radio waves, which are the world's chief message-bearers, travel in straight lines; the world, however, is round. Television programmes, which operate on very short waves, are not held in by the ionosphere. They go straight out into space and are lost. Transmissions across the Atlantic for example, instead of homing on the other side, leave the world's surface at a tangent and are gone for good. To keep them where they are needed, the television

Sir Bernard Lovell, head of the Radio Astronomy Department of Manchester University and 'father' of the radio-telescope at Jodrell Bank, talks to reporters in a corridor of the station on the night of the first successful soft landing on the moon.

engineer can do two things. He can set up a series of sea-borne repeater stations—floating versions of the microwave links at present used on land. These, placed at thirty-mile intervals across the Atlantic, could pass the signal from one to another, keeping it well down on the surface. But this method, more logical than imaginative, is scarcely practical.

The other course is even less practical: this would be to build a mast in the middle of the Atlantic, its reflector unit high enough to pick up and send on straight-line transmissions from both shores. The disadvantage here is that, to be effective, the mast would have to be nearly five hundred miles high.

A third possibility is to reflect signals from the surface of objects in orbit. This idea, 'science fiction' in the 1940s, was by the 1950s a practical possibility. Experiments had already been carried out from one object in orbit: the moon. In a series of trials culminating

The historic picture of the moon's surface (left above), transmitted from Russia's Luna 9 and 'developed' at Jodrell Bank with the aid of newspaper facsimile, was a world scoop. It was, however, distributed to the press of the world by the 'Daily Express' without fee. The picture above shows some of the world's newspapers on the following morning.

in 1959 speech echoes were obtained at the naval research laboratory in the United States and then by Jodrell Bank.

Later, instead of bouncing signals back to their starting point, Jodrell Bank set up a joint experiment with the Commonwealth Scientific and Industrial Research Organization in Australia. Early in 1961 teleprinter messages went from Jodrell Bank to Sydney by way of the surface of the moon.

The moon was by no means ideal as a reflector, but specially constructed satellites would be. When in 1957 the Russians had successfully put up the world's first artificial satellite it became immediately clear that the ideal was not far off.

Clarke's idea, pre-dating the *Sputnik* by more than ten years, called for something even better. This was a satellite system, a set of three satellites in fact, each circling the earth at a distance of some twenty thousand miles, each moving at a speed exactly

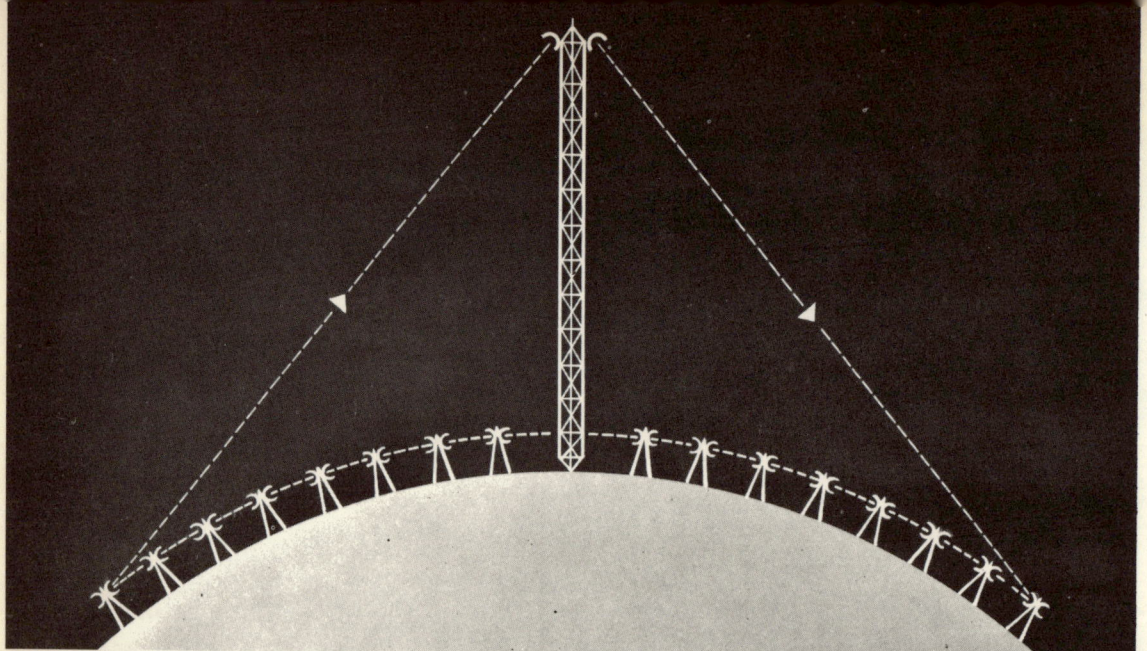

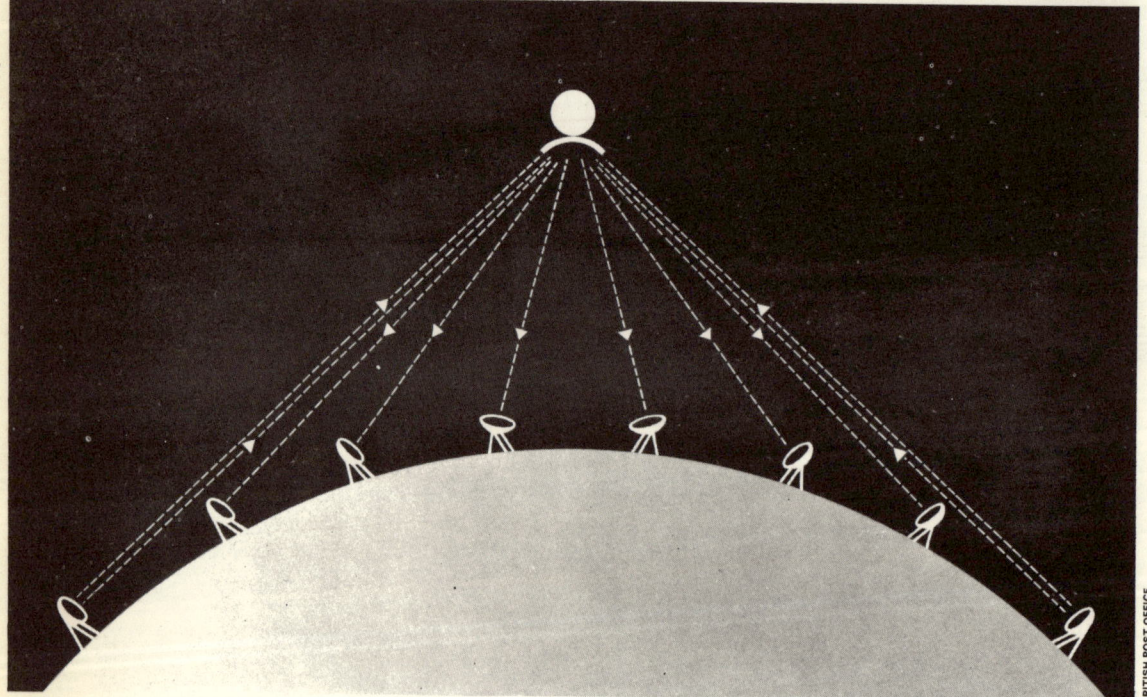

corresponding to the turning movement of the earth. With this arrangement, each satellite would cover approximately one third of the world's surface, and each would appear, relative to any spot on earth, as a fixed point in space.

The first communications satellites, *Telstar*, *Relay*, *Earlybird* and others, had the disadvantage of high speed and low orbit. This meant that their useful 'bounce' duration was short; they had passed through their period of mutual visibility within twenty minutes or half an hour. It also meant that the bounce equipment, aimed from the ground, had a difficult job in keeping the satellite fixed in its sights, and, of course, transmissions were restricted to specific orbit times. Useful though they were, their value was limited.

Within a matter of a few years satellites were launched exactly as Clarke had suggested. These, actively re-transmitting instead of merely reflecting, used the sun as a source of power. Circling at some twenty-two thousand miles from the earth, they appear virtually stationary and are operational all round the clock. Their specialized radio characteristics allow them to be used for many more than one purpose at a time; each has a potential capacity of many thousands of telephone calls—all being transmitted simultaneously and each arriving at its destination as a separate and distinct signal.

The story of Arthur Clarke's vision is, historically speaking, a short one. Hardly had he expressed the idea than it became a fact. It was an idea that in a decade has transformed the world's concepts of communications. 'Science fiction' as it used to be called has become everyday science fact—as familiar as the TV screen and the telephone.

Stationary or otherwise, the communications satellite is the most potent communications medium ever conceived. People living in the huge territories of Russia, China, Africa and India, previously beyond the reach of television, are suddenly well within the world arena. From a few central points, in many cases just one, audiences of many millions may be reached.

The Soviet *Orbita* system is typical: rather than set up multiple microwave repeaters in remote regions, Russian TV engineers have produced large numbers of steerable earth stations; centred on appropriate areas, these 'lock on' to *Molniya* satellites and pick up TV transmission direct from the Television Centre at Moscow. Other countries are beginning to share in the network. During the 1970s *Orbita* ground-stations are scheduled to operate in Cuba, Africa, the Middle East, South East Asia and Latin America.

India's first communications satellite earth station goes into action in the 1970s too. This, India's share in the Communications Satellite Operations CONSAT operation, will utilize up to one hundred and thirty of the twelve hundred channels available on the stationary *Intelsat 3* satellite. India's TV plans provide for the progressive installation of community TV receivers in each of the country's five hundred thousand villages.

In Alaska three ground stations are to transmit educational and general information programmes through NASA's Applications Technology Satellite. TV transmissions, specially tailored for the needs of people living in remote areas, will total some seven hours a day. The satellite, *ATS I*, is in stationary orbit over Hawaii. A similar satellite is in stationary orbit over South America.

This is the pattern of communications in the 1970s. It is safe to say that by the middle of the 1980s much, if not most, of what people see and hear will come to them from far above their heads.

Educational television, brought by satellite to remote but densely populated areas, is transforming life and attitudes in many countries. This United Nations picture shows a TV lesson in Nigeria.

Diagrams on the opposite page show the basic facts of satellite communication. Radio waves travel in straight lines; whereas ordinary radio waves are held in to the curvature of the earth's surface by the enclosing ionosphere, shortwave transmissions go off at a tangent. To keep them on the surface, radio engineers could build either a five-hundred-mile-high radio mast with reflectors at the top (above), or a series of 'short-bounce' floating stations in the Atlantic. Using a stationary satelite however, (below) signals can be reflected over a very wide area. Three such satellites, properly spaced, can cover virtually the whole of the world's surface.

77

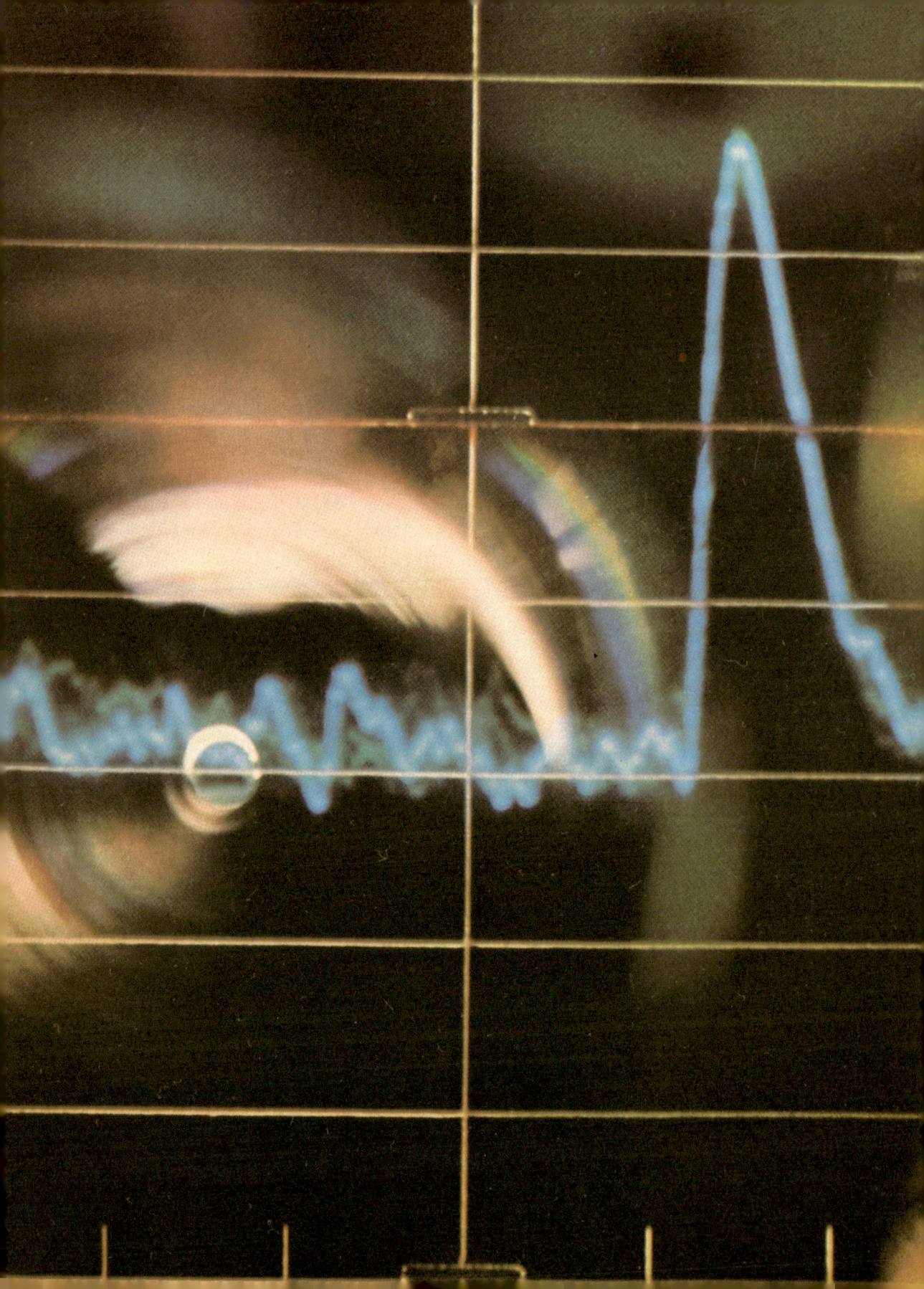

INTO THE EIGHTIES

New Problems—New Solutions

COMMUNICATION, so lately geared to the plodding pace of man and horse, now flickers round the globe like a sheath of summer lightning—all-pervading, instantaneous and unbroken. It is as though the layer of atmosphere in which we breathe and move has now become charged with a new ingredient. As well as oxygen, nitrogen and the other components of life-support, there is now Communication . . .

It is a development full of portent. In scarcely more than a generation the tempo and volume of keeping in touch has moved up half a dozen notches at one time; it has gone straight from Very Slow to Instantaneous, from Scarcity to Overdose. For millions of younger people the build-up of communications is no novelty. For their elders too, some of them born hardly out of the stagecoach era, its acceptance has become almost as ready. Mankind is quick to adapt.

With its enormous signal-load potential, the satellite has brought new dimensions to human contact. But already communications engineers foresee saturation point: early satellites were hardly bigger than a beach ball; in the 1970s they are developing into ten-, twelve- and fourteen-foot giants, with circuit capacities enormously increased. In the foreseeable future space techniques will allow of even more massive structures. Size is no problem; it will ultimately be possible to put into orbit installations of virtually any size—whole 'buildings', if necessary. To operate effectively these will require to be powered not from today's solar cell units, but from atomic energy. This also presents no problem; atomic-energy power units are already successfully at work in implanted pacemakers in human heart surgery. Tomorrow's satellite, like much of tomorrow's technology, will have all the power it needs.

What it will be short of, however, is 'radio-frequency room'. The problem, as with frequency-apportionment in ordinary radio, is to keep messages from overlapping each other on their way to the recipient. When frequencies are too close, when too many transmissions jostle for radio space, quality suffers. Even with today's advanced transmission methods, in which signals are 'shuttled' from one channel to another as they become momentarily vacant,

AMERICAN TELEPHONE

At the Bell Telephone Laboratories a scientist studies key elements of a millimetre waveguide system. In the late 1970s the principle may be used to transmit up to two hundred and fifty thousand messages simultaneously.

Increasing congestion in the St Lawrence has prompted the Canadian Department of Transport to set up an electronic birds-eye information system covering some two hundred and fifty miles of the river. Seated at consoles before a thirty-three foot electronic wall map, operators keep track of all vessels as their presence is reported on VHF radio; a computer estimates speeds and courses, and warns of possible collisions.

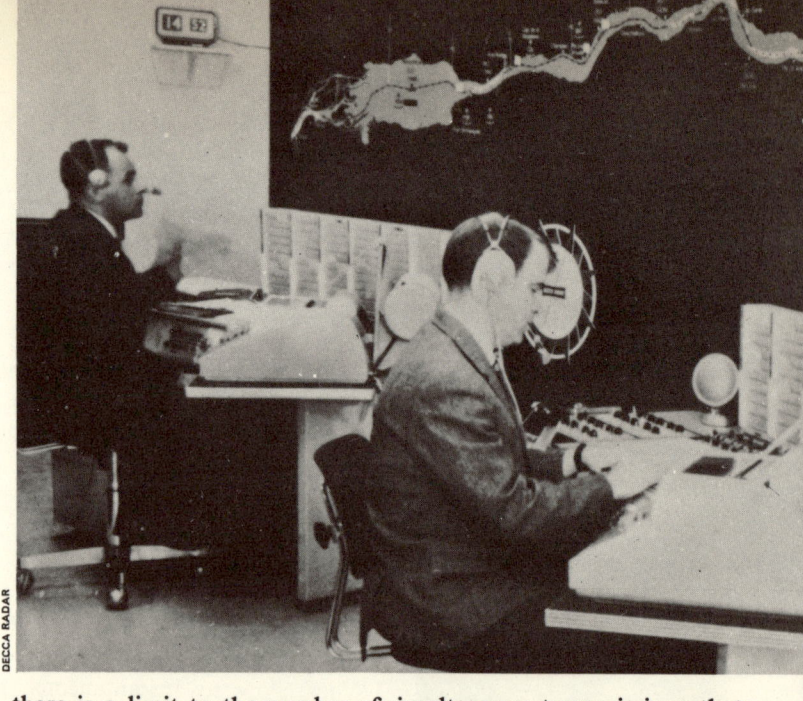

Micro-miniaturization of printed matter allows over three thousand pages of ordinary library books to be recorded on a transparency not much bigger than a playing card. The system is revolutionizing information storage technology. The picture below shows micro viewers in use at the Directory Enquiry section of the Hamburg telephone service. The picture on the right compares the bulk of a set of encyclopaedias with a micro-version of the same work contained in the transparencies in the man's hand.

there is a limit to the number of simultaneous transmissions that even the satellite can handle.

There is another problem: though radio signals travel at the speed of light (186,000 miles per second) their arrival at a distant point is necessarily slightly later than the moment of transmission. Thus, when radio messages were bounced experimentally from the surface of the moon, it took two and a half seconds for the echoed signal to get back to earth. Greater distances would produce correspondingly longer delays, but even at the comparatively short distance of twenty-two thousand miles (the orbit of a stationary

satellite) there is a time lag of roughly one sixth of a second. This means that *hallo?* at point A takes twice that time to go via the satellite to point B, and that an answering *hallo!* from B will not reach A for another third of a second. The round trip, even if the distant speaker answers immediately, takes two thirds of a second.

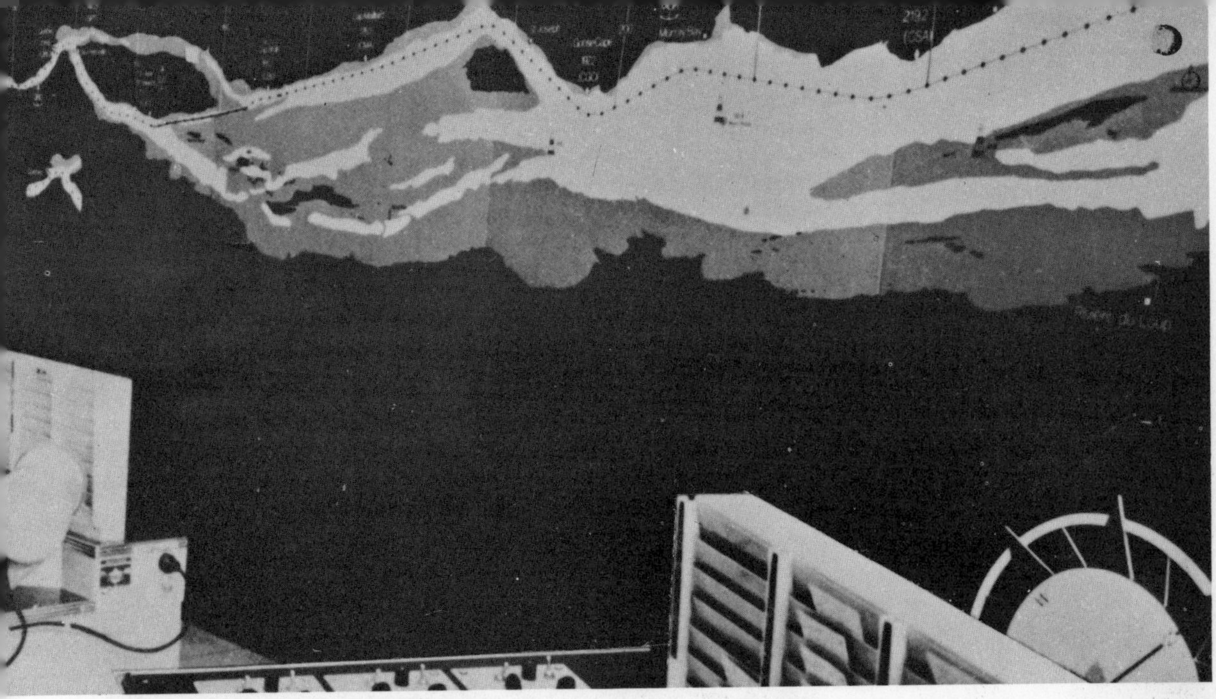

For ordinary conversational purposes, this time-lag can be disconcerting. People react differently to the delay; some fail to notice it, others find it frustrating. In the give-and-take of everyday conversation, most people respond to speech not after they have heard it, but while it is still coming over; whether with actual words or only by occasional sounds of acknowledgment, they interrupt or 'overspeak' each other. Only rarely does one speaker complete the last syllable of a sentence before the other starts.

The delayed action response of the satellite circuit is a problem that communications engineers have learnt to live with. For the general public, a time lag of two-thirds of a second is the maximum acceptable delay. With any longer period normal conversation becomes impossible.

In multiple satellite systems, where two or more are used in a round-the-world hook-up, the delay problem becomes crucial. Here, because of the increased 'bounce times' involved, conversational exchange is out of the question. The circuit can only be used for one-way transmission—of television programmes or other information signals where a few seconds delay is unimportant.

Even in one-way transmission the delay factor cannot be overlooked entirely: to save channel space, some television transmissions send vision by satellite and sound by cable—which offers no noticeable delay; to ensure that the separate signals arrive simultaneously, engineers are obliged to introduce an appropriately timed artificial delay in the cable transmission.

Radio contact is a feature of public transport in many of the world's big cities. In the picture a traffic monitor controls London buses from a central console.

The satellite, though ideal for some communications jobs, has its limitations. For two-way person-to-person contact the future may lie where it started—back in the cable.

Cable capacity was initially limited by the tendency of a long-distance signal to fade in strength. The difficulty was overcome by installing 'repeaters' in the cable. These served to boost the signal on its way, passing it on to the next point with renewed strength. The technique has been brought to a fine pitch; transistors have taken over from valves in the repeater unit, giving greatly increased reliability and much reduced power demand.

Traffic control operators at Schiphol Airport supervize aircraft arrivals and departures from computerized consoles. Plans for all flights are fed into the computer; the system coordinates all data, from basic flight plans to approach-risk detection.

Combined with improvements in transmission methods, progress in cable technology has begun to offer loading performance to match the satellite. In only ten years, cable capacity has increased thirty- to forty-fold. The first repeatered Atlantic cables carried thirty-six circuits; tomorrow's North Sea cables will carry thirteen hundred circuits. By the early 1980s a single submarine cable is expected to carry three to four thousand circuits. The cable has additional attractions: whereas a satellite message costs as much for ten miles as for ten thousand miles, the cost of a cable signal is commensurate with the cable's length; whereas satellite message capacity has limits (and transmission technology is already in sight of these) the new-style cable has virtually none.

Among the biggest contributors to the new-style cable concept has been the technique called Pulse Code Modulation. PCM was invented by Alec Reeves as long ago as 1937, but for the full realization of its potential it had to wait for the transistors and integrated circuits of the fifties and sixties. Essentially, PCM consists of transmitting not the complete signal but *samples* of it. For transmitted speech the effect is similar to that of the movie film on the eye. Just as the cine camera records only selected samples of what is going on (twenty four samples per second) PCM chops up sound into small portions. As with film, where the gaps between the samples are filled in by the brain of the viewer, PCM transmission conveys to the listener the illusion of continuous speech. The number and speed of frames in PCM—eight thousand per second—is enormously high compared with movie film, but there is still plenty of room, electronically speaking, to *fit in other signals in the gaps between them.*

It is as though a movie film, as well as having the normal twenty four frames per second, had *frames between the frames*—pictures forming another movie altogether and which, by some optical trick, were invisible to one section of the audience while the pictures of the main film were invisible to the other. In this way there would be two simultaneous but distinct channels of communication on the same reel of film.

In PCM speech transmission, instead of there being just one extra channel, there is room for very many more. By the early 1970s telephone engineers had crammed in from twenty-four to thirty-two separate conversations on one pair of wires. More were to come. Experimental work showed potential signal loads to be virtually unlimited. The technique requires only that individual impulses be made shorter and shorter to allow progressively more room for others.

A cable concept of a different sort is the microwave waveguide. Here again, mid-century technologies have culminated in a radically new approach. Using wide-band signals shorter in wavelength even than the short waves used in commercial television transmission, engineers now send signals from point to point through pipes. Guiding and containing its contents like a water pipe, the hollow waveguide takes carrier waves to their destinations, each carrier bearing as many as eighteen hundred signal channels.

Waveguides are of two kinds: rectangular and circular. The rectangular cross-section is cheaper to manufacture and install, but offers high resistance to its microwave content; the cylindrical, or circular, type offers less resistance but requires precision finish and allows only minimal curvature. Circular waveguides giving point-to-point transmission over long distances open up tremendous possibilities. Waveguide engineers visualize capacities of *up to one hundred thousand simultaneous telephone conversations through a pipeline only five centimetres in diameter.*

Yet another pipeline concept—fibre optics—has come to the fore in the seventies. Glass fibres, each hardly thicker than a human hair, have the property of transmitting light along their length with remarkably little loss on the way. What is much more signifi-

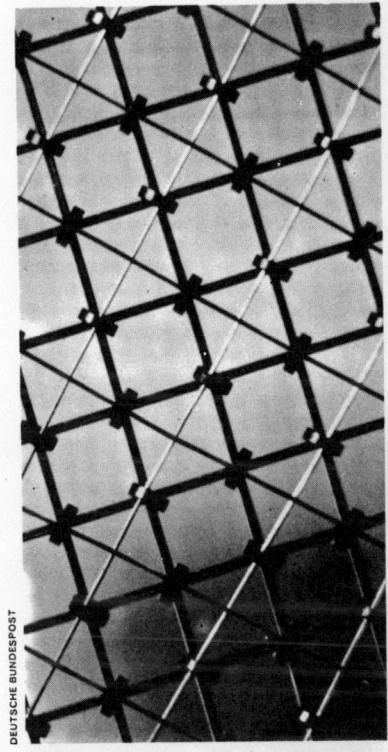

Computerization of information systems has gone hand-in-hand with the development of 'solid-state' miniaturized electronics. This ring-core storage system, used in the research department of the German Post and Telegraphs, channels information in thousandths of a second.

cant, though the fibre strand may twist and curve, the light passing through it is retained inside, emerging at the other end with only slightly diminished intensity. As with the micro-waveguide, the effect is similar to that of a gas or water pipe. The principle is widely used in medical instrumentation; clusters of fibres unite to form inspection probes for examination of the interior of the body. Each fibre provides a micro-lense unit in an overall mosaic picture, transmitting what it 'sees' to the outside world.

For the communications engineer the potential is obvious, here again is a duct through which signals may pass—a communications channel for tomorrow.

Side-by-side with the development of fibre optics has come another development—the laser. This, a system of highly concentrated light-transmission, projects a beam so intense that it can travel to the moon and back without losing itself in the atmosphere on the journey. Whereas an ordinary light beam 'scatters', laser light remains intensely focused—a fine pencil of brilliance all the way. As a new pathway for communications, the laser had obvious potential from the start. Communications engineers realized that, given the right conditions, the beam could be used as a carrier of impulses—a signal-sender.

Allied to fibre optics, the laser became even more interesting. Used on its own, the laser, like the flashing mirrors of the ancient

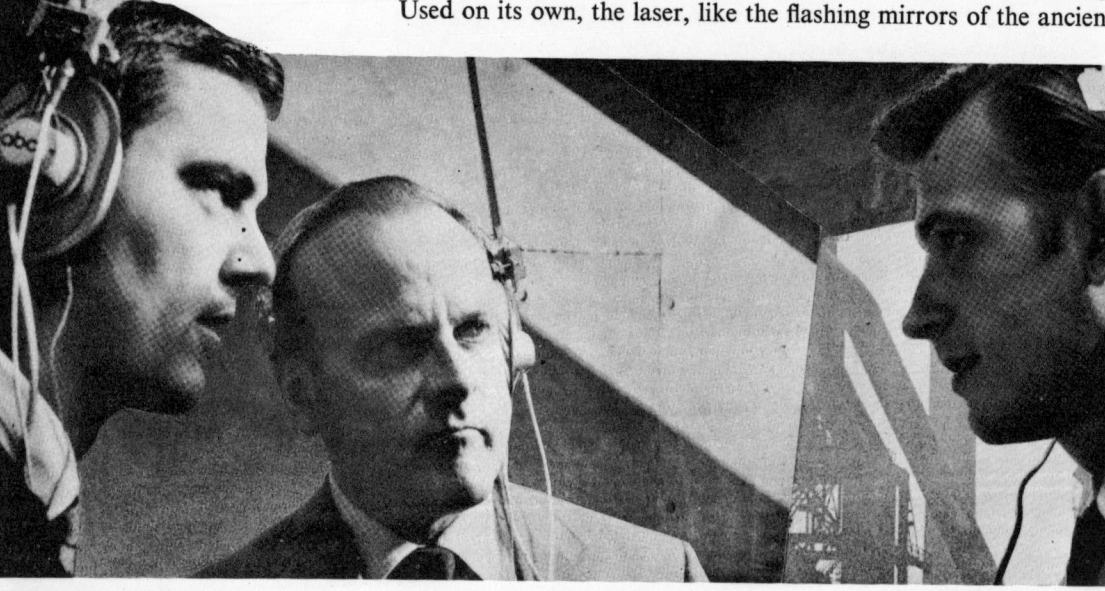

At major news events 'floor reporters' keep contact with radio transmission studios and may also feed in live interviews on cue from central control.

Greeks, was at the mercy of atmospheric visibility. But used in conjunction with a glass fibre cable, it could be exploited to the full—channelled and diverted this way and that at will.

There are technical difficulties; the fibres, if they are to transmit light efficiently, have to be of exceptionally pure glass. In addition, each fibre requires a 'core' of glass of another kind. There is also the unattractive problem of jointing and splicing. On present showing, say communications engineers, the job is more appropriate

JAPAN INFORMATION CENTRE

ST DUNSTAN'S

AMPEX

Communications technology has left few aspects of daily life untouched. The blind man in the picture above is using an 'invisible walking stick' to avoid obstacles. A hand-held ultrasonic scanner detects obstructions and converts their presence into audible warning signals. Criminal records (left, above), until recently bordering on the limits of physical manageability, are now being taped and stored on micro-unit information spools. The system allows huge reductions in storage space and search time. In the lower picture American TV newsmen operate hand-held mobile television cameras. Independent of cable links or mains supplies, cameras can operate in black and white or colour, sending pictures to central transmitters or to videotape for recording—or both.

Telecommunication first made an appearance in civil aviation on board a KLM Fokker on its first regular flight from Schiphol, Holland, to London in 1922. The equipment was in the form of a radiotelephone and telegraph-receiver. Today Schiphol airport is one of the world's most advanced centres of aviation communications technology. The picture shows the scene in part of the central control tower as controllers monitor aircraft movements on the ground. Airfield radar (in the centre of the console) shows a complete birds-eye view of the whole airport; operating at a very high rate of antennae rotation, the radar picture is renewed five hundred times a minute, enabling images of fast-moving aircraft to be registered during landing and take-off. Radar is a major item in airport control equipment; at Schiphol facilities include long-distance radar, which observes the entire area of the Amsterdam Flight Information Region and also approach radar, with a shorter range. Multiple display units allow selective viewing; duplication of vital equipment is built in on both systems. In emergency each can take over the other's functions.

The waveguide—a hollow tube through which pass signal-carrying microwave impulses—functions almost in the same way as a gas or water pipe. It serves as a containing channel, conveying its contents where they are needed. To minimise power-loss, waveguides are machined to a very high degree of precision. There are two basic types, circular and rectangular in section. The circular type, shown here, offers only very slight resistance to power passing through it, but cannot easily be curved. Rectangular waveguides dissipate a lot of their power but allow greater variety of conformation. Waveguide technology permits transmission of large quantities of information simultaneously.

to a neurological surgeon than a telephone lineman.

But the rewards are obvious. Using PCM techniques, as now applied in the conventional cable, a single glass fibre has been shown to be capable of carrying two hundred thousand telephone conversations at once. International cables, converted to fibre optics, could carry literally millions of simultaneous calls.

The communications network of tomorrow is a complex pattern of inter-operating systems: to satellites and line-of-sight microwave transmitters will be added new waveguide networks and new 'glass cable' links. Superimposed upon the whole will be increasingly refined techniques of Pulse Code Modulation with greater and greater load capacity all the way round. The vision is one of almost unimaginable scale and complexity.

In this new labyrinth of interconnection where will the ordinary man stand? Will the present urge for contact prevail, or will Mankind suddenly decide it has had enough? There are signs already that some would prefer a little isolation.

There are signs too, of the new burdens that tomorrow's communications will bring; the TV telephone brings its own problems; so does the long-distance wrong number; so does the

POSTES ET TELECOMMUNICATIONS FRANCAISES

computerized credit-status memory bank; so does the instant electronic referendum centre. Communication is not an unmixed blessing; there are many who view it as a growing menace.

The TV telephone will give way to the 'wristwatch' TV telephone; the national insurance number will give way to an all-purpose 'citizen number', electronically processed and updated in a National Data Centre; remote-controlled surveillance cameras will give way to invisible remote-controlled surveillance cameras . . . These, and many other side-effects of the communications explosion, will eventually touch the life of everybody on our planet.

In August 1970 it was reported from Louisville, Kentucky, that the United States Defence Department was taking a close look at a proposed emergency warning system which would permit the President of the United States to contact ninety-five per cent of the American people within sixty seconds. Carl D. Russel, head of the group that had filed a patent for the system, explained that by dialling a code sequence on a special telephone instrument the President could 'take control from the White House of all radio and television stations plus every telephone in America'. The system, said Mr Russel, would automatically turn on radios at full volume and would activate broadcast stations that were off the air.

It takes little or no imagination to see the implications of Mr Russel's idea and to extend its principles to other appliances. The same is true of a score of communication devices: the 'bugging' gadget, the implanted electrode, the long-distance alarm—all are forms of communication, all are infinitely versatile—and all are subject to misuse.

In the early 1970s one British manufacturer was making a car-thief deterrent. The device is as small as a ballpoint pen and is carried by the motorist when he leaves his car. Although miles away, the device flashes a warning to the owner when the car's ignition is switched on. But the device does not leave it at that; it offers two-way communication: by pressing an attachment on the 'pen' the owner transmits a coded impulse to his car—immobilizing the engine.

In the late 1960s neurologists in the research laboratories of an American university fitted an electrode into the brain of a monkey. Transmitting signals to the electrode from a distance by radio, it was possible to control the animal's eating behaviour. By moving a switch to the left the animal could be made to eat ravenously. Moving it to the right produced vomiting. Alternating between the two positions, the changes could be repeated indefinitely.

Towards the end of 1970 the same university reported two-way radio communication between an animal's brain—and a computer. Electrodes implanted in a chimpanzee's brain picked up electrical activity in the cortex and sent signals to the computer, which in turn sent back signals to another part of the brain, correcting and modifying the animal's behaviour. The Yale University team were reported as saying that the methods suggested 'promising new ways

A tubular system of an earlier kind appears in the Paris 'Pneumatic Post'. The system, which has counterparts in a number of cities, dates back to the turn of the century. Messages are contained in cartridges which are blown, or sucked, at high speeds along the tubes. As with the circular waveguide, there can be no sharp turn, deflection of the tubing must be in smooth curves. The Paris system extends under the city for some four hundred kilometres. With the advent of hovercraft and 'air-cushion railway' techniques, communications scientists of the seventies are taking another look at this method.

of treating mental and physical disorder in humans'.

Dr Gray Walter, of the Burden Neurological Research Institute in Bristol, is reported to have transmitted a radio impulse triggered by electrical activity in the human brain. By merely *thinking*, the experimenter was able to send a signal—a signal which could be used to switch a machine on or off, or to perform a number of other simple functions...

It begins to look as though Mankind may soon be facing a

There are now some hundreds of satellite earth stations in various parts of the world. This one, at Pleumeur-Bodou in France, is of the 'ear-trumpet' variety and is enclosed in a spherical radome. The radome, though it offers no obstruction to radio waves, keeps out wind and weather. Among many functions Pleumeur-Bodou also operates in conjunction with Goonhilly Down in Britain, Raisting in Germany and Fucino in Italy. It can also carry colour television from and into the Soviet Union's 'Molnya' system.

communications overdose. But there is another problem too, less daunting, though nonetheless real. As communications techniques improve, man's mobility decreases. The millions who in the thirties, forties and fifties went to the cinema, no longer do. Instead, the pictures that they see are carried into their own homes on the shortwave frequencies of television. In the big urban centres all over the world, stay-at-home viewers see the world from wherever they happen to be. Without moving from the spot, Man in the communications age sees and hears more of his planet than even the most privileged of his predecessors. Rooted firmly to one spot, he ranges far and wide.

As travel has become easier, cheaper, quicker and safer, it has become strangely less exciting. For the man of the seventies there is no virtue in movement for its own sake. Movement from place to place has become, from its very familiarity, devalued. We may expect that by the same token, as communication becomes every day more matter-of-fact, this too will become devalued. When long-distance telephone calls were a wild extravagance people took them seriously, and planned what they had to say. Today the long-distance call often consists only of chit-chat. As Donald Wray, Deputy Director of Engineering of the British Post Office has expressed it, 'when a letter to somebody in the antipodes took six weeks to get there, and cost a lot of money to send, people took trouble over letter writing. Even the short-distance letter was quite an event; take Lord Chesterfield's letters to his son for example. But you can be sure that in the eighties and nineties there will be no bound volumes of telephone conversations between somebody and his nephew in Ontario...'

AMPEX

RANK PRECISION INDUSTRIES

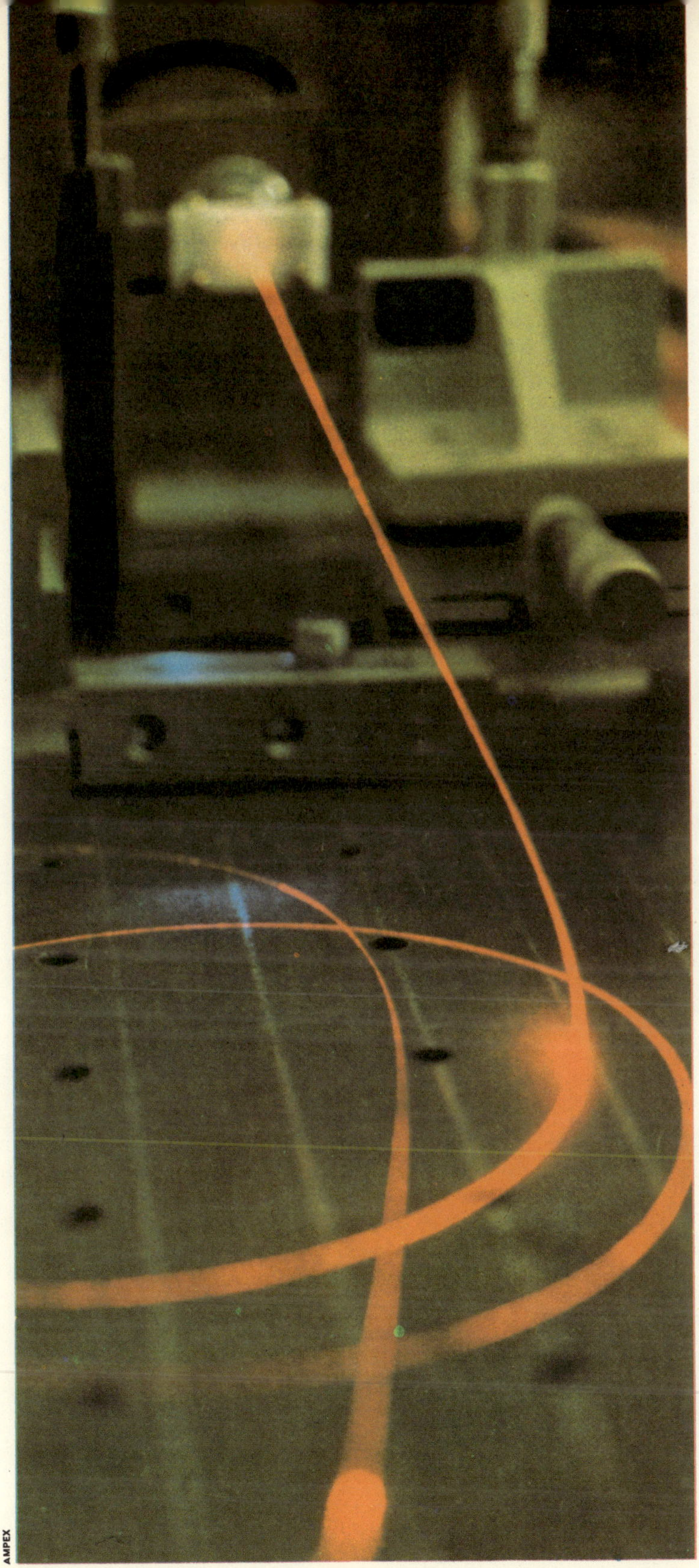

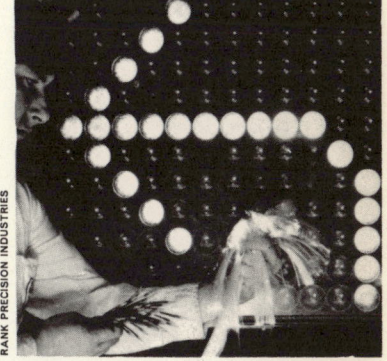

Glass fibres—tiny filaments as fine as human hair and smaller—are among the media now being developed for message transmission. As the experimental set-piece in the larger picture shows, 'glass cables' have a capacity for carrying light over distances and round corners. Communications scientists are exploiting this property for a wide range of applications. Fibre optics are already in use for 'seeing-round-corners'. The circular cells in the variable traffic sign in the smaller picture are fed with light conveyed through flexible light-pipes. The technician is seen holding typical glass fibres in one hand and light-pipes in the other.

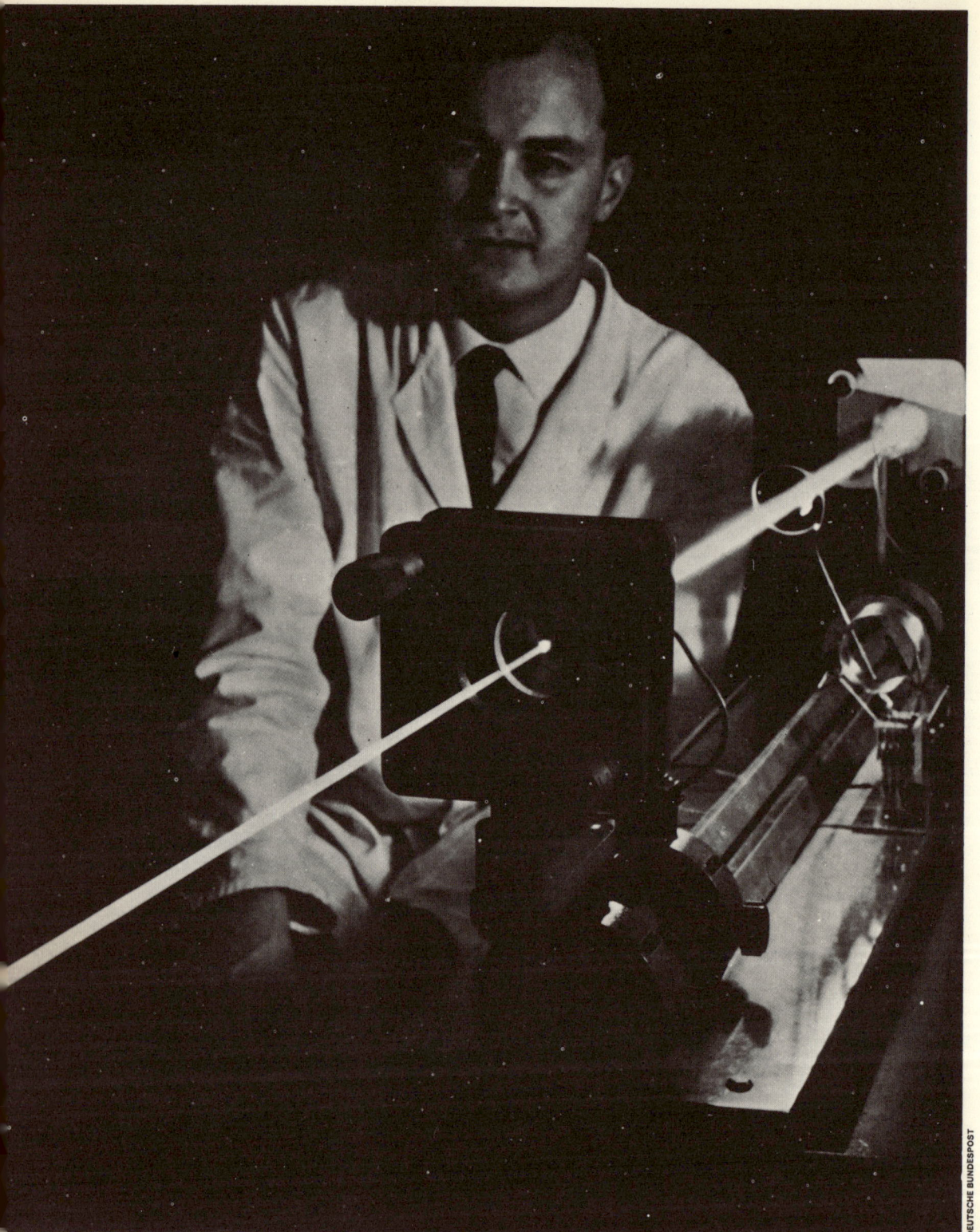

The growing network of communication has one dimension as yet scarcely touched. Signals between the planets, messages through interstellar space, the whole concept of contact with worlds beyond our own—this is a dimension so new to Man and so unexpected, it has taken him almost by surprise.

The first practical lesson has been the delay effect. Signals bounced from the earth to the moon and back again take time to travel. The further they go the longer they take; a conversation between a man on earth and a man on Mars would require six-minute silences between question and answer.

Between earth and Pluto, farthest of our solar neighbours, the time-lag would be ten hours—almost a whole day for an answer to come back. But Pluto, though distant in our immediate neighbourhood, is only a step away compared with places outside our system. Conversational exchanges with the nearest star would require gaps of eight and a half years.

It is clear that interstellar space which, it must be remembered, is many millions of times greater than interplanetary space—offers little in the way of casual chat. Whoever may be out there would need more than a passing interest in earth to commit himself to an extended conversation. (With something in the region of one thousand million million million other solar systems to choose from, we seem an unlikely choice.)

The time-lag of interstellar space is a barrier as great as space itself. But science does not ignore the proposition that messages may be being sent. Dr D. M. A. Mercer, the British physicist, told a meeting of the British Association for the Advancement of Science that there may be some hundred million places in our universe where conditions are suitable for the development of life. That meant, he said, 'a possible hundred million advanced communities.' He went on: 'Within fifteen light years of us there are seven "suitable" stars; within fifty light years there are one hundred "suitable" stars. There is a small probability that some advance communities are within hailing distance. They may already be in a "galactic club", busy communicating; they may maintain a small sub-department who are beaming messages at other likely stars in the hope that they may discover some novice community, just on the threshold of interstellar communication, to whom they can send their kindergarten messages and gradually educate to their standards . . .'

Dr Mercer went on to say that there is already a listening station in the United States which has searched—so far without result—for intelligible radio waves from nearby stars.

There is no reason to think that this planet, with its life-supporting atmosphere, is unique among countless millions. Whatever the barriers, whatever the difficulties, it is likely that in the next few decades Man will learn to receive and send his kindergarten messages. The sheath of communication that circles our planet may soon put out tendrils, to link with others.

The laser beam (opposite page), here seen in the research laboratories of the German Post Office, has the property of 'coherence'. Whereas an ordinary light beam suffers loss of strength through 'scatter', the laser beam remains densely concentrated and compact. This means that it can cover very long distances (for example, to the moon and back) without serious loss. By introducing modulations into the beam it can be made to convey information. Combined with glass-fibre technology and new modulation engineering, the laser is expected to solve many of today's problems of cable and satellite overload.

As well as light, the laser provides a concentrated beam of heat. It is used as an aid in ophthalmic surgery in a 'spot-welding' process for the re-attachment of detached retina. Because it is released in pulses of only one thousandth of a second, the patient feels no sensation of pain.

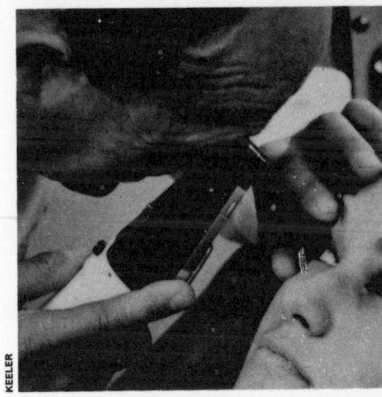

Index

Numerals in Italic indicate illustrations

aircraft, 13, 28, 82, 86
American Civil War, 44
Asahi Shimbun, 61, 62
The Australian, 62

Baird, John Logie, 54, 55
beacons, *7*, 8
Bell, Alexander, 22
Brunel, 19, *22*

cables, 18–19, *22*, *36*, 38, 43, 49, 73, 81–83
cathode ray tube, 55
Clarke, Arthur C., 73, 75, 77
coaches, *11*, *12*, 28
computers, 41, *47*, 49, 57, *59*, 69, 80, 82, 83, 89
consoles, *50*, 51, *57*, 80, 82, *86*
Crippen, Dr, 16–18, 25

Daily Express, 62, *66*, 72, 75
Defoe, *11*
disc television system, 55

Eidophor screen, *59*
electricity, 15, *16*, 18, *19*, 23, 24, 26, *42*, 80, 89, 90
electrophone, 23

facing letters, 29–30, 32, 38
facsimile machines, *61*, 62, *63*, *65*, *66*, 70, 72
fibre optics, 83–84, *91*
Field, Cyrus W., 19

Great Eastern, 19, *22*

Havas, Charles, 46
Hill, Rowland, 23–25, 26
hot-air balloons, 13–15, 18

Jodrell Bank, 70, 73, 75

Kendall, Captain, 16

laser, 84, *92*, *93*
Le Sage, George Louis, *9*
letters, *14*, 23, 25, 29–32
Lincoln, Abraham, 44
Lovell, Sir Bernard, 70, *73*

Macadam, *11*
mails, 12, 14, 23, 25, 26, 28, 35, 67

Marconi, Guglielmo, 18, *24*
Marconi-Emi Electronic system, 55
Mars, 73, 93
messages, 7, 8–9, 25, *47*, 49, 50, 79, 89
messengers, 8–9, 15
microfilm, *14*, 15
micro-miniaturization, 15, 80
microwave transmission, 42, *43*, 69
moon, 70–73, 74, *75*, 80, 93
Morse, Samuel F. D., *18*, 20
morse code, *18*, *24*
Mount Pleasant Sorting Office, *28*, *31*, 32

news, 13, 44–51, 57, *63*, 70, 84
News Agencies, 44–51
newspapers, 7, *8*, 14, *15*, 44, 61–63, 66, *72*
Night mail, 32

Odler, Louis, 15

pack horse, *11*
parcels, 23, 26, 28, 29, *32*, 40
penny post, 23, 24, 26
photography, 13, 15
Pony Express, 9, 13, 28
post, 14–15, 23–25, 26, 28, 67
postage stamps, 24–25
postal code, 38–39, 42
postal conference, 1863, 25
postal handling, 26, 28–43
postmen, 32, 39
Post Office railway, *35*
Post Office tower, 26, *43*
pigeons, *8*, 13–15, 44, 51
Pulse Code Modulation, 82–83, 88

radar, *86*
radio, 18, 61, 62, 63, 65, 72, 73, 77, 79, 80, 81, 84, 86, 87, 90, 93
radiotelescopes, 70, 72
railways, 13, 18, 28
Reuter, Julius, 44–49, 51
roads, *11*

satellites, *59*, 62, 73, 75–77, 79, 81, 82, *90*
semaphore, *15*, 18
Siege of Paris, 13–15, 18
space travel, 56, 70
speech, 7, 9, 23, 26, 36, 41, 59, 73

Surgical University Clinic of Munich, 52, 56

tape, 67, 68, 69
telegraph, 13, 18, *20*, 22, *23*, 25, *38*, 44, 51
telephone, 3, *16*, *19*, *20*, 23, 26, *39*, *40*–43, 49, 58, *65*, 73, 77, 83, 86, 88, 89
telephone boxes, *19*
teleprinter, *3*, 26, 49, 50, *51*, 62, 75
television, 7, 26, *43*, 52–59, 62, 67–69, 73, 77, 81, 88, 89
telex, *3*, 42, 62
Titanic, 44, *46*
traffic, 56, 57, *81*, 82
train, *12*, 30, 44
Tresuguet, 11
turnpikes, *11*
typewriters, 42, 49, 50

Universal Postal Union, 25

video, 61, *69*, 85
Volta, Alessandro, 15

waveguides, 83–84, *88*
Western Union, *20*
wireless, 16, 18, *25*
Wolff, Hermann, 46

Xerxes, 7

Acknowledgements

The author and publishers would like to express their thanks for help, guidance and picture material given in the preparation of this book by Donald Wray, Deputy Director of Engineering of the British Post Office; Jack Lush and John Auld of the Press Association; Doon Campbell and Clifford Wakefield of Reuters; David Oates and Jan C. Geel of NV Philips' Telecommunicatie Industrie, and William Newton, Telecommunications Manager of the *Daily Express*.

Thanks and acknowledgement are also due to the following for further help and picture material:

American Telephone and Telegraph Company

Ampex International

Australian News and Information Bureau

British Post Office

Bundesministeriums für das Post- und Fernmeldewesen

Christopher Elliott

Creed and Company

Daily Express

Decca Radar Limited

EMI Electronics Limited

Federal Communications Commission

Ferranti Limited

Japan Information Centre

Johanna Harrison

Keeler Optical Products

Imperial War Museum

International Telecommunication Union

Ministère des Postes et Télécommunications

Modern Telephones Limited

Muirhead Limited

National Cash Register Company

National Physical Laboratory

NV Philips' Telecommunicatie Industrie

Press Association Limited

Pye of Cambridge Limited

Rank Precision Industries

Reuters Limited

St Dunstan's

World Meteorological Organization